When Men and Mountains Meet

Great things are done when men and mountains meet;
This is not done by jostling in the street.
WILLIAM BLAKE. *Gnomic Verses*

WHEN MEN AND MOUNTAINS MEET

The Explorers of the Western Himalayas
1820–75

JOHN KEAY

KARACHI
OXFORD UNIVERSITY PRESS
OXFORD   NEW YORK   DELHI
1993

Oxford University Press, Walton Street, Oxford OX2 6DP
OXFORD NEW YORK TORONTO
DELHI BOMBAY CALCUTTA MADRAS KARACHI
KUALA LUMPUR SINGAPORE HONG KONG TOKYO
NAIROBI DAR ES SALAAM CAPE TOWN
MELBOURNE AUCKLAND MADRID
and associated companies in
BERLIN IBADAN

OXFORD is a trade mark of Oxford University Press

First published by John Murray (Publishers) Ltd., U.K. 1977
This edition in Oxford Paperbacks, 1993

ISBN 0 19 577465 5

Printed in Pakistan at
PanGraphics (Pvt) Ltd., Islamabad.
Published by
Oxford University Press
5-Bangalore Town, Sharae Faisal
P.O. Box 13033, Karachi-75350, Pakistan.

For
Christina, Kranti
and Rahul

Acknowledgements

In a book based on works long out of print and on archive material the author's principal debt must be to the libraries. I should like to acknowledge the unfailing assistance and advice of the staffs of the India Office Library and Records, the Royal Geographical Society, the London Library and the National Library of Scotland. Valuable suggestions on specific queries have been made by Colonel Gerald Morgan, Dr. G. J. Alder, R. G. Searight and H. D'O. Vigne; to all of them I am most grateful.

The Scottish Arts Council helped make the project possible, numerous friends have provided encouragement and in particular Christina Noble, John Murray and Roger Hudson have read the full text and made comments both kind and helpful.

Lastly my thanks to Julia. She has done far more than just type the whole thing; but how to acknowledge the patience, assistance and encouragement of a collaborator and wife?

Contents

Maps

Illustrations

ENGRAVINGS

ACKNOWLEDGEMENTS

Nos. 1, 17, 18, 19 and 23 are reproduced by kind permission of Christina
Noble; No. 2 by courtesy of Mrs. Lee Shuttleworth and Penelope Chet-
wode; No. 3 from *Joseph Wolff* by H. P. Palmer, London 1935; No. 4 from
Victor Jacquemont published by the American Philosophical Society, Phila-
delphia 1960; Nos. 5, 13, 14, 15, 16 and 20 by courtesy of the Royal Geogra-
phical Society; No. 6 by courtesy of H. D'O. Vigne Esq.; Nos. 7 and 8 from
Travels in Kashmir by G. T. Vigne; No. 9 from *A Short Walk in the Hindu
Kush* by Eric Newby, published by Hodder & Stoughton; Nos. 10 and 12 by
courtesy of the Director of the India Office Library and Records; No. 11
from *A Personal Narrative of a Visit to Ghuzni, Kabul and Afghanistan* by
G. T. Vigne, published by Whittaker; No. 21 from *Lahore to Yarkand* by
George Henderson, London 1875; No. 22 by courtesy of Gerald Morgan
(photo by Ney Elias) and No. 24 by courtesy of Nick Holt.
    The engravings are all taken from *Travels in Ladakh* by H. D. Torrens,
London 1862.

Introduction

Arab geographers of the Middle Ages, though they knew more than most about the configuration of the world, entertained some rather fanciful notions. One of these was to regard the Eurasian landmass as a desirable woman clothed in nothing but a long chain girdle about her ample waist. The girdle was of mountains studded with snowy peaks. It stretched from the Pyrenees through the Alps, Balkans, Caucasus and Elburz to the limits of the known world in the Hindu Kush and Himalayas. This idea of an east–west mountain range encircling the earth was not new; the geographers of Greece and Rome had also subscribed to it. But it would have taken the genius of the Arabs, had they known the region, to recognise in the Western Himalayas the girdle's jewel-encrusted clasp. Nowhere on the earth's surface is there a comparable cluster of mountains. In a chaos of contours at the heart, or perhaps the navel, of Asia six major mountain systems lie locked together.

From the south-west, out of the arid hills of Afghanistan, comes the Hindu Kush. Its brown, treeless slopes grow higher and steeper as they approach the main complex. In the valleys there are no strips of continuous cultivation, just lush green oases strung together by grey, snow-fed rivers. Even below the glaciers of the 25,000 foot Tirich Mir there is still a hint of the deserts of the Middle East. The inhabitants in their small dusty villages are all, with the one exception of the Kafirs, Mohammedans. Skull cap and turban have their mountain equivalent in the rolled caps of Chitral and Gilgit, but the henna-red beard is still in evidence and *salaam aleikum* is the eternal greeting. From a miniature white-washed mosque the muezzin goes up into the thin mountain air and rouses the ragged goatherd on the hills above.

To the north the Hindu Kush is bounded by the narrow Oxus valley beyond which lie the rounded domes of the Pamirs. This is a still mysterious and almost polar region of several parallel ranges linked together, and to the other chains, by a north–south range which includes the peak of Muztagh Ata, 'the father of snowy mountains'. The Arabs called the Pamirs the *Bam-i-Dunya*, or 'Roof of the

World', a fitting name for a region where the valleys are not cosy little clefts but open steppes as cold and windswept and almost as high as the peaks that bound them. Here, in summer, graze the yaks, horses and sheep of the nomadic Kirghiz, a Central Asian tribe of squat gnarled men and smiling flat-faced women. Conditions are so harsh that they are said to suffer from the highest incidence of still-born babies in the world; and in winter even they forsake the mountains to shelter in the lower valleys to the north. The Pamirs are left to their one notable resident, the Marco Polo sheep, a creature as big as a pony with curled horns up to six feet long.

Corresponding to the Hindu Kush but joining the Pamirs from the east is the third system, the Kun Lun. The Pamirs with their Siberian climate are a fitting contribution from Russia, but the Kun Lun belongs in every sense to China. Rich in jade and overlooking the ancient silk route from Cathay, its row of blue peaks emerging above the dust haze to the south of Kotan have always been regarded as the rim of the Chinese empire in the south-west. The Turki population constitutes a sort of Celtic fringe, acquiescing in the rule of Pekin, occasionally rejecting it but never really finding a workable alternative.

Within the angle formed by the junction of these three systems lie the Karakorams. Mightiest of all, they radiate from an amphitheatre of peaks that includes three of the world's six highest. This is a treacherous and perpendicular wilderness where only the mountaineer can hope to survive. Glaciers thirty miles long and quarter of a mile deep fill the valleys. Anything like level ground is strewn with their moraines. Grazing, let alone cultivation, is practically non-existent; no one lives actually in the Karakorams.

By comparison the Great Himalaya,* in its western reach, is a modest affair. The part which falls within the Western Himalayas is that which is drained by the Indus and its tributaries. It is more broken than on its grand sweep through Nepal, with passes as low as the 11,000 feet Zoji La between Kashmir and Ladakh. In Lahul and

---

* It is important to distinguish between Himalaya, pronounced like 'Somalia', and the Himalayas, as in 'Malaya'. The first is, strictly, the correct pronunciation, and deriving from two Sanscrit words meaning 'the abode of snow' should be employed in the singular only. I have used it only when referring to a specific mountain chain as in Great Himalaya. Himalayas, always plural, is an Anglicisation which is now more widely used than the correct version. I have taken it in its broadest sense to imply all the mountain systems between the Indian subcontinent and Central Asia.

Zaskar and along the Indus valley there is a sizeable population. Their villages are not clusters of low huts sheltering in the floor of the valley, but one or two square and many-storied houses rising boldly out of the hillside. With flat roofs aclutter and faded clothes— or are they prayer flags?—hanging out the windows, they look like a poor example of hasty urban re-housing. But inside, life is more peaceful. In all but political allegiance Ladakh is part of Tibet. The people are Buddhists, many of them monks or nuns. In the still air the squeak of a prayer wheel is answered by the single resounding stroke of a gong which issues, pregnant with meaning, from an unseen monastery in the rock face overhead. A charming world, one thinks, this Great Himalaya of Ladakh. Yet the culmination of the range in the Nanga Parbat Massif at the heart of the whole mountain complex is a staggering contrast and a worthy climax. Rising almost sheer from the bed of the Indus at 3,000 feet to some 26,000 feet, it boasts such a host of towering ridges, precipices and peaks that it is often regarded as a range in itself.

Finally there is the Pir Panjal. Lying between the plains of India and the Great Himalaya this, too, steals some of the latter's thunder. By Himalayan standards it is not high; there are few peaks over 20,000 feet. But, because it rises fairly abruptly from the plains and is the first snow-capped range encountered by the traveller from the south, it deserves separate note. Anywhere else in the world it would surely be rated a noble range. As it is, the Pir Panjal is simply one of the loveliest in the world. For there is a beauty here that is neither savage nor tame. Compared to the grim and barren ranges to the north, all is green and bountiful. The grass is deep and dewy, the forests of oak, sycamore, pine and cedar are of truly Himalayan proportions and the timbered houses, roofed with uncut slates, nestle in hollows white with apple blossom. Spring seems ever present in the Pir Panjal. Yet the snowfall here is heavier than anywhere else in the Western Himalayas. The peaks look nearer, higher, less assailable. The passes are every bit as treacherous as those of the other systems and the gradients just as painfully steep.

The area covered by this complex of mountains is nearly the size of France, in shape a broad belt about six hundred miles long by three hundred wide. It was this three hundred from south to north which confused and dismayed all those who tried to penetrate the mountains. For these six systems are not just ranges. Each has its own network of spurs and subsidiary parallel ranges. Some of them, like the Ladakh range south of the Karakorams and the Aghil range north of them, are as formidable as their better-known neighbours. Nearly all trend roughly from east to west. A journey across this axis from Punjab to Turkestan involves climbing fifteen major passes, all between nine and nineteen thousand feet high. Excluding the small vale of Kashmir and some elevated Tibetan tundra in the east, there is no level ground between the mountains.

A jet today will cross the whole region in less than an hour. It is assuredly the smallest of the theatres of nineteenth century exploration. Yet it ranked with, if not above, all others including Africa, Australia and the Polar regions. It took as long to penetrate, it entailed as much hardship and was invested with as much importance. The awards presented by the Royal Geographical Society in London provide as good a yardstick as any. For journeys made between 1831, when the Society was founded, and the end of the century, twelve of the pioneers of the Western Himalayas received the Society's Gold Medal. No other comparable corner of the globe can equal this record. And this in spite of the fact that the Society was never as closely involved in Himalayan exploration as it was, for instance, in African.

Why then is the story of the Western Himalayas so little remembered today? Emphatically it is not because it was a parochial Anglo-Indian affair. It wasn't. French, German, American and Russian travellers all contributed. The geographical societies of Europe responded eagerly to every advance, and the governments of Britain and Russia watched developments with a rapt attention bestowed on no other region of the then unknown.

Nor was it because the mountains were in any sense less exciting. There were no naked savages, it is true, but romance was almost as good as titillation and of this the Himalayan lands had more than their share.

> Who has not heard of the Vale of Cashmere
>   With its roses the brightest that earth ever gave
> Its temples, and grottoes, and fountains as clear
>   As the love-lighted eyes that hang over their wave?

Lalla Rookh, An Oriental Romance by Thomas Moore was written in 1826. Drawing heavily on the rich sentiments of the Persian poets, and on the few accounts of Kashmir then available, it epitomised existing notions of this most famous of the Himalayan valleys. Moore had never been to Kashmir but his balmy scent-laden breezes rustling the bowers so dear to lovers somehow managed to conjure up a not unrecognisable image of the place. Kashmir was, and to some extent still is, a sensuous paradise. And if the first visitors were a little disappointed not to find more love-lighted eyes it was, as they conceded, only because the Kashmiri girls were so lovely they were more commonly met with in the zenanas of Lahore.

Moore left his heroine, Lalla Rookh, happily married to the man she loved and tripping off 'over the snowy hills to Bucharia'. Bucharia, or Lesser Bucharia, was one of the many names for Eastern Turkestan, now Sinkiang. One cannot help fearing for the safety of the delicate jewel of the East. She sets off so gaily, quite unaware that ahead lie five of the toughest passes in the Himalayas along what was later called 'the ugliest track in the whole wide world'. Moore and his contemporaries had some idea of Kashmir but none at all of the mountains beyond it. In fact it was mid-century before either their size or extent was appreciated. Cartographers of the day were still influenced by their classical counterparts. Rennel's map of 1794 shows Ptolemy's long skinny lines of peaks, the Imaus and Emodus, streaming across wide open spaces where dwell the tribes of Comedi and Byltae. Herodotus had populated the area with unicorns and gold-digging ants as big as foxes. And the prodigious wild sheep of the Pamirs, which Marco Polo reported, were still as challenging a notion as that other Himalayan speciality, the yeti.

In so far as the maps gave no real idea of the terrain, travellers were less concerned with exploring the mountains than with crossing them. Bucharia, Kashgaria, Moghulistan and Tartary were all names for the fabled lands beyond. Not much was known of their geography, but what was known was the existence there of some of the oldest and richest cities in the world. Bukhara and Samarkand, Kashgar and Yarkand, Kotan and Khokand, these were places that fired the imagination. They lay along the ancient land-locked trade routes from China to the Middle East, the routes by which the silk and the porcelain of Cathay had reached the Mediterranean. For centuries they had been shrouded in mystery, forbidden but not forgotten. About 1270 Marco Polo and others of his family found them recovered from the ravages of Genghis Khan and as prosperous as ever. Samarkand was 'a great and noble city', Kashgar famous for its estates and vineyards, Kotan for its gold and jade, and Yarkand 'where they have plenty of everything' even had a Christian bishop.

In the nineteenth century the rediscovery of these places constituted one of the greatest challenges. The unknown of darkest Africa had its appeal, but the slightly known of Central Asia was thought by many to be every bit as exciting and rewarding. It appealed not just to the missionary, the merchant and the naturalist but also to the antiquarian, the scholar and the statesman. The explorers of the Western Himalayas, though as ill-assorted a crowd as can be imagined, have a few traits in common. They tend to be more scholarly than their counterparts elsewhere, more wide-ranging in their interests and, perhaps because of the nature of the mountains, more chastened by their experiences and more reticent about them. There are few lions amongst them but a lot of elusive and intriguing chimeras.

By way of the low Hindu Kush passes in Afghanistan, Western Turkestan with its twin cities of Bukhara and Samarkand was soon reached. Eastern Turkestan proved far more difficult. The easier route via Afghanistan and the Pamirs had to be abandoned because of political difficulties with the Afghans. The more direct approach north from Kashmir taxed the patience and credulity of a host of pioneers. No sooner was one range negotiated than another loomed up ahead. The passes grew higher and higher, the track became a nightmare. Not till the late 1860s did travellers at last return with news of Kotan, Yarkand and Kashgar. They were not exactly a disappointment but their remoteness was now at last fully appreciated. Those with a commercial or political interest in Eastern Turkestan tended to minimise the difficulties of the track, but there was no disguising the fact that the barrier between India and Central Asia was infinitely more formidable than was at first thought.

In other words the mountains were coming to overshadow the lands that lay beyond; they were now seen as in themselves constituting a legitimate field for exploration. This change is traceable from about 1860 when the survey of the whole region got under way. Earlier travellers, like Vigne and Thomson, who had gone down as Central Asian explorers who failed to get there, were resurrected as eminently successful Himalayan pioneers. People began to get some idea of the undreamt of breadth and complexity of the mountain knot. And those long skinny chains of peaks slowly gave way to something more like the vast purple bruise, riven with muddy browns and spattered with icy whites, of a modern contoured map.

The exploration of the Western Himalayas had therefore this retrospective, almost incidental character. The early travellers described at length the peculiarities of the Himalayan peoples and the ruggedness of the terrain but, until the mountains were seen as a unique region in their own right rather than as a simple barrier, only qualified acknowledgement greeted their achievements. This perhaps is one reason why today the story is so little understood. Another must be the fact that in Asia exploration seldom led on to colonisation. Explorers talked of making Kashmir 'a little England in the heart of Asia', of 'opening up Central Asia' and of 'bagging the Pamirs'. But none of these things came about. These were not virgin tracts for the land-hungry settler; the inhabitants were neither uncivilised nor helpless. Moreover the commercial and agricultural prospects that might lead to colonisation simply did not exist. The fertile valleys of the mountains and the oases of Turkestan were already intensively cultivated and were too inaccessible for exploitation. As usual the terrain was the deciding factor. To this day thousands of tons of fruit rot each year in the valleys of the Western Himalayas for want of a means of exporting them.

Then, as now, the position was further complicated by political barriers. The Himalayas are a great natural frontier, and a frontier is like a tidal beach; a zone of instability within which the loyalties of the scattered population can ebb and flow with the tides of power outside. What the British Indian empire, with its notions of imperial defence and centralised administration, wanted was a boundary, that is, a line not a zone, a sea-wall not a beach. And there is no more difficult, less 'natural' boundary than a 300-mile-wide knot of mountains.

The frontier aspect of the region is well shown by the diversity of the peoples who inhabit it. Interlocking with, rather than confronting one another, the worlds of Mongol, Aryan and Turanian here coalesce. This ethnic jigsaw has its religious counterpart with Budhist, Hindu and Islamic adherents occupying neighbouring valleys. Polygamy is the rule in one place, polyandry in another. Here the wildlife is sacred, there it is hounded unmercifully. For the boundary maker this ethnological confusion only compounded the geographical problems. It is a measure of the complexity of the human and physical geography that the boundaries of the Western Himalayas are still today in dispute. No less than five countries, Pakistan, Afghanistan, Russia, China and India, now have a toe-hold in the mountains.

In most parts of the world the explorations of traders, missionaries and geographers paved the way for political penetration. In the Western Himalayas it was more the other way round. From 1840, if not earlier, political considerations dictated the pace and direction of exploration. This not only made life very difficult for the un-official traveller but also shrouded in secrecy the efforts of those who had official blessing. Time and again one feels that the published accounts of an expedition tell only half the tale. The other half can sometimes be tracked down in the government records of the period. But not everything was committed to paper and the bona fide traveller is often hard to distinguish from the political spy.

In the last thirty years of the century when the direction of ex-ploration swung away from Yarkand towards Gilgit and the west, this air of secrecy and intrigue deepened. This is another story* but suffice it here to note that as the Western Himalayas came into their own as a recognised field for the explorer, so the atmosphere grew more highly charged and the circumstances of individual journeys became more obscure. It all adds another fascinating dimension to the story but it also explains why so few attempts have been made to reconstruct it.

*   *   *

The extant accounts of the region by pre-nineteenth century travellers were not much help to the early explorers. In many cases it was the achievements of the explorers which inspired the scholars to unearth them rather than the other way round. The travels of Buddhist monks in the fifth to seventh centuries AD from China to India in search of their religious heritage were not public knowledge till the 1860s. With their strange tales of dragons and haunted lakes they would not have been much help anyway; geographers are still unable to decide on the precise routes which they describe. The next rele-vant account was that of Marco Polo's crossing of the Pamirs from Balkh to Kashgar in 1274. This was at least well known and most of the nineteenth century travellers owned a copy. The great Venetian merely skirted the real mountain complex and to this whole leg of the journey devotes only three or four pages. His account, ever cele-brated for its obscurities and bizarre details, had not been greatly reverenced in the Age of Reason and it took the explorations of the nineteenth century to reinstate its credibility.

* An account of 'The Gilgit Game' is currently being prepared by the same author.

The only people who had really penetrated the Western Himalayas before 1800 were Jesuit missionaries in the two preceding centuries. To the casual observer, Buddhist ritual has unmistakable similarities to that of Rome. On the strength of reports that the Tibetans were as good as Catholics already, the Portuguese Jesuits in Goa pushed missionaries into Western Tibet and even to Lhasa. In the course of this work Fr. Francisco de Azeveda in 1631 penetrated as far as Leh in Ladakh, and in 1715 Fr. Hippolyte Desideri en route from Kashmir to Lhasa also reached Ladakh. Another Jesuit, the lay brother Benedict de Goes, crossed from India to China by way of Kabul and Marco Polo's route across the Pamirs. He died of exhaustion soon after reaching China but his account was pieced together by the Jesuits and was certainly known in the early nineteenth century. Desideri's achievement was also recorded, but the full account of his journey was not edited till 1904, while Azeveda's was only made public in 1924. However the fruits of their labours were incorporated in the map produced by d'Anville in 1735 and this in turn was used by Rennel. Thus, very indirectly, the early explorers of the nineteenth century owed something, if only their knowledge of the existence and position of Ladakh, to the Jesuits.

There were also two, rather different, British sources on the area and one by a Frenchman. In 1808, in anticipation of a possible Napoleonic repeat of Alexander the Great's march, the British government in India despatched an imposing mission to Kabul. It was commanded by the Hon. Mountstuart Elphinstone and had a trained surveyor in Lieutenant MacCartney. His area of personal observation did not include the Western Himalayas, but using native informants he formed a better idea of the mountain area than any of his predecessors; it was incorporated on the map of the mission. At last all the classical names disappear. For the first time the Upper Indus is given a westerly flow, and a range called 'Mooztagh or Kurrakooram' makes its cartographical debut north of the 'Hemalleh' mountains. There are a lot of mistakes, a lot of empty spaces and tidy, random ranges, and a hopeless muddle over proportions and locations. But the nomenclature and hydrography are recognisable. It would be no help to the explorer when planning his route, but it was at least capable of correction.

In all this, exception must be made for Kashmir. This valley is far larger, more prosperous, more populous and more accessible than any other in the Western Himalayas. It lies between the Pir Panjal and Great Himalaya, enjoys a perfect climate and is famed as much for the charms of its women as for the quiet beauty of its landscape and the workmanship of its craftsmen. Here there was more than enough to attract the attention of merchants, travellers and conquerors. The Moghul emperors turned the place into a giant pleasure ground, and it was in Aurangzeb's entourage that a Frenchman, François Bernier, visited the valley in 1663. Like so many subsequent visitors, he set about analysing its attractions and somehow failed quite to capture its seductive appeal. But, from his detailed and enthusiastic account, the Western world divined a hint at least of the glamour and excitement that the mere mention of Kashmir conjures up in the oriental mind.

In 1783 George Forster, an Englishman, was the next to try his hand at describing the valley. He too failed, though it was hardly surprising; most of his short stay was spent trying to find a disguise that would see him safely out of the place. He was en route from Bengal to England overland, a remarkable journey which he successfully completed by way of Kabul, the Caspian and St. Petersberg. He arrived in Kashmir disguised, as he thought, as a long lost native of Turkestan. Unable to conceal his identity in so simple a matter as urinating from a crouching rather than a standing position, he was easily faulted. In Srinagar, the capital, he was compelled to go into hiding and eventually left as an Armenian trader. Forster's travels added little to the existing knowledge of either Kashmir or the Western Himalayas. The most that can be said for them is that they are highly entertaining and, in this, considerably ahead of their time.

It was left to Moore to capture something of the voluptuous undertones of Kashmir. *Lalla Rookh*, along with Marsden's edition of Marco Polo and Elphinstone's *Caubul* were the three books which all the early Himalayan travellers possessed. An ill-assorted armoury, they offered more of inspiration than practical assistance. Professor H. H. Wilson in his preface to the travels of William Moorcroft summed up the situation.

> . . . of the countries upon their [the British possessions in India's] confines we know less than we do of the central deserts of Africa. The whole of the intervening country between India and China is a blank; and of that which separates India from Russia, the knowledge which we possess is but in a very slight degree the result of

modern European scholarship and is for the most part either un-
authentic or obsolete.

This was written in 1840, some fifteen years after Moorcroft's great
journey; its careful phrasing was partly necessitated by discoveries
made in that intervening period. In 1820, when Moorcroft pioneered
the exploration of the lands between India and Russia, no more was
known of them than of what Wilson calls the blank between India
and China. Neither, of course, was a complete blank but there was
precious little that was known for sure.

NOTE: Even today knowledge of the geography of the Western
Himalayas is not something that can be taken for granted. There is
no handy coastline to fix their location in the mind and in most
atlasses they are cut about so that the southern ranges fall within a
map of the Indian subcontinent, the western ones within a map of
Russia and so on. A few minutes study of the accompanying maps is
recommended.

PART ONE

The Pir Panjal and The Great Himalaya
1820–1838

I'll sing thee songs of Araby
And tales of wild Cashmere,
Wild tales to cheat thee of a sigh
Or charm thee to a tear.
W. G. WILLS. *Lalla Rookh*

India and Central Asia showing the approximate positions of the Rus

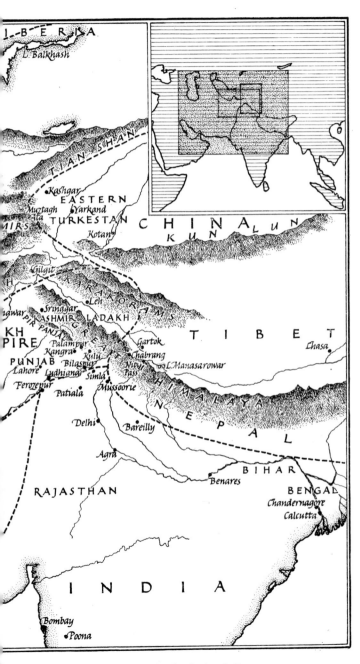

Indian frontiers and the intervening lands, in 1838

1. The Enthusiastic Moorcroft

Buffalo skins on the Sutlej

The river Sutlej, longest and most southerly of the Indus tributaries, cuts deeply into the rim of the Tibetan plateau, thunders in a dark sinuous gorge through the Great Himalaya and then slows to an unexpected standstill below the small town of Bilaspur. Its current is stayed by the towering dam of the Bhakra hydro-electric project.

Before the river was harnessed it was here about fifty yards wide and swept strongly past at five miles an hour. On the morning of March 6th 1820 William Moorcroft measured it while his party of three hundred men, sixteen horses and mules and several tons of baggage was being ferried across. The operation took an hour and a half, good going considering there were no boats. Everything had to be transported on inflated buffalo skins—or rather on the backs of the paddlers of the inflated buffalo skins. Moorcroft watched it all carefully. The skins were blown up through one of the legs. A thin rope was tied round the body and, grasping this with one hand and a paddle with the other, the ferryman flopped chest down on to the upturned carcase and awaited his load. The Englishman was impressed. Ships could be equipped with these as life rafts. A few buffalo skins could render a whole army amphibious. In a waterlogged land like Bengal or in the Punjab, should they ever have to fight there, this could be of real advantage to the East India Company's forces.

It is typical of Moorcroft that in his enthusiasm for the skins he forgets to note in his diary that, crossing the Sutlej, he is bidding farewell to the last outpost of British territory in north-west India. He is also entering the Western Himalayas. For a more romantic spirit it would have been a stirring moment. Grasping the shoulders of the half-submerged paddler and pushing off into the cold green waters was a step into the unknown, the beginning of a marathon journey which was always mysterious and ultimately tragic. It was to establish its leader as the father of modern exploration in both the Western Himalayas and Central Asia.

Considering he left behind him ten thousand sheets of manuscript scattered through Asia but now safely in London's India Office Library, we know pitifully little about William Moorcroft. Few travellers have written at greater length or on a wider variety of subjects and yet remained so personally obscure. To some extent this was convention. Marco Polo had managed to relate the story of his twenty-five years of travel without revealing more about himself than a passion for hunting. With both men one suspects a desire to be known more for their achievements than their character. But to the would-be biographer it is most discouraging and, for lack of one, Moorcroft's reputation has suffered. He still gets described as a horse-dealer, a spy or an adventurer, each of which possesses a grain of truth while sadly understating his real standing.

Not that there were any obvious signs of greatness about the man who was watching his party safely across the Sutlej. He was short and slight, red faced and fair haired, and he favoured one of those nautical peaked caps which Dr. Livingstone would invariably wear. Contemporaries speak of his 'candid manner' and 'energetic disposition' but it is by no means clear whether these are to be regarded as virtues. That Moorcroft liked to speak his mind in spite of, and usually regardless of, the consequences is self evident from his writings. And that he had an energetic disposition is not surprising; loafers make poor explorers. From one of his companions it appears that he was also dubbed 'an enthusiast', and this is important. In those days 'an enthusiast' was not a flattering description. It implied imbalance, a man with convictions as far-fetched as they were unshakable. In this light the 'energetic disposition' suggests excitability as much as wanderlust, the 'candid manner' bloody-mindedness as much as integrity.

Given his career—and it is here, as he would have wished, that his greatness lies—Moorcroft's character may best be seen as approaching that of the erratic genius. He was indeed obstinate, eccentric and controversial, but he was also full of infectious energy and invariably ahead of the times. To anyone responsible for his activities he was an unholy terror but to his friends and followers a guiding light. He should have been safely cloistered in a progressive university rather than let loose on the Himalayas. Empires are created by men of vision but Moorcroft was a visionary.

The one thing that might have struck anyone watching this odd little Englishman as be bestrode his buffalo skin was his apparent age. From dates in his early career that can be established Moorcroft must have been born about 1765. In other words the man now starting on one of the longest and most dangerous journeys in Asian exploration was fifty-five years old. By the standards of the day that made him an old man. In terms of life expectancy amongst the British in India it was positively ancient. One stands aghast at his extraordinary courage and stamina. In the whole story of Himalayan exploration there is nothing to compare with it; almost without exception his successors were men in their twenties and thirties.

At the same time it must be remembered that this makes him very much a product of the eighteenth century. Though he was to foresee the whole course of events in the Himalayas and Central Asia and though his own travels fall well within the nineteenth century, his attitudes are those of an earlier age. His models are Marco Polo and the Tibetan traveller, George Bogle. Unlike the Victorians he never notices the Almighty's role in the wonders of nature. Nor does he subscribe to that simple equation which would make progress equal civilisation and civilisation equal Christianity. If he praises British rule it is because, compared to the lawless and extortionate administrations he encountered, it seemed fairer. Christianity he seldom mentions, but he shows a real respect for other men's religions, especially Buddhism.

The eighteenth century with its free-thinking and its commercial drive was also the founding period of the British Raj. In the last quarter of the century, under Governors-General like Warren Hastings, Cornwallis and Wellesley, the East India Company had developed from being just one of several European trading companies with a stake in the country to being the greatest power on the subcontinent and the administrator of an Empire which dwarfed the combined possessions of the British crown. By 1820 the map of peninsular India looked much as it would for the next 150 years. However this rapid growth had been achieved at great cost to the

finances of the Company. It was time to call a halt to territorial
acquisition and to concentrate on more profitable activities such as
revenue settlements and curbing expenditure. Like the Directors
of the Honourable Company, the British Government also was
cautious about expansion. They already exercised a degree of control over the Company and they were increasingly aware of their
responsibilities in India. Moorcroft too probably had little sympathy
with those who advocated more conquests; but he was still, at heart,
a founder. Cautious retrenchment was not for him. He travels not
exactly flag in hand but with many a backward glance as to its likely
progress. The obsessions which filled his hungry mind and the mis-
understandings which dogged his footsteps were as much a product
of his unfashionable outlook as of his character.

One must also see this, his last journey, as a do or die effort. His
career is virtually over, his young family is being sent back to
Europe* and it is clear that if he returns he intends to resign from
the East India Company's service. He continually reverts to the
question of his retirement and, in the face of criticism and dis-
couragement, he can afford a certain detachment. It also helps explain
the complete lack of urgency in his movements. He is not out to make
his name or his fortune. When his merchandise won't sell, his
political overtures are disowned and his resources exhausted, there
is still every reason for going on. It is his final fling. He stakes everything, including his life.

As a young man he had abandoned medical school in Liverpool and
had gone to France to study veterinary surgery. This was in the late
1780s, when the science was virtually unknown in Britain. He returned as the first qualified vet in the country, set up practice from
an address in Oxford Street in London and quickly made a small

---

* Moorcroft's family affairs are as confused as most things about him.
News of his wife's death reached him in Ladakh in 1820. Her name was
Mary Moorcroft and she had died in Paris. But this was evidently not the
mother of his two young children. Anne, the elder, who was staying with
Mary Moorcroft at the time of her death, describes her as neither as fair nor
as young nor as pretty as her own mother, who was in India. One would
suspect that Moorcroft had taken an Indian bride except that in that case he
would hardly have sent Anne to Paris and England or have planned to send
the little boy, Richard, to Harrow and medical school. Equally, from a lock
of long fair hair belonging to Mary Moorcroft which is preserved amongst
the Moorcroft papers, it would seem that the second Mrs. Moorcroft must
have been very fair indeed.

fortune. He became joint professor of the Royal Veterinary College, founded in 1791, and wrote such esoteric pieces as 'Directions for using the Portable Horse-Medicine Chest' and 'Experiments in Animal Electricity'. A turning point came with the attempt to market his patent machine-made horseshoes. To the relief of village smithies and the eternal benefit of geographical science the venture was a flop. The small fortune was lost. Moorcroft gratefully accepted an offer from the East India Company and in 1808 set sail for India. He was to manage the Company's stud farm in Bihar, a highly responsible and lucrative appointment in those days of cavalry.

Much was expected from the supervision of such a distinguished man of science, and at first Moorcroft did not disappoint. Ever a pioneer, he introduced oats to India and by 1820 could claim that for every ten sick animals in his care when he arrived there was now scarcely one. The only thing that bothered him was the quality of his horses. He badly needed new breeding stock. Scouring India, or 'running over the country in quest of phantoms' as it seemed to his patrons, he found nothing suitable. Either a herd of English stallions would have to be shipped out to Calcutta or he would go in search of the horses of Central Asia, famed since the days of Marco Polo for their speed and endurance. The English stallions were not forthcoming. Nor were the Company enthusiastic about their stud superintendent disappearing into what was then a great void in the maps of Asia. But he was not expressly forbidden to go and for Moorcroft this was enough. He had a pretext.

That the purchase of horses, particularly the Turkoman steeds he expected to find at Balkh and Bukhara, was not to be more than the official justification for his travels is clear from an early date. Moorcroft's vision was all embracing and as much commercial and political as scientific. In 1812, on the first of his two famous journeys, he had crossed the central Himalayas into Tibet disguised as a Hindu *saddhu* and gone in search of the goats from which the fine wool of the Kashmir shawl is extracted. With Hyder Young Hearsey, an Anglo-Indian adventurer, he had scaled the Niti pass in Garhwal, just west of the present-day frontier of Nepal, and reached the Tibetan centre of Gartok and the sacred lakes of Manasarowar and Rakas Tal at the source of the Sutlej. To some geographers this journey is his greatest achievement. He was not, as has been claimed, the first European to cross the Himalayas; the Jesuit missionaries in the seventeenth century had used an adjacent pass to reach their mission centre in

Chabrang on the Upper Sutlej. Nor was he the first Englishman; George Bogle had crossed the Eastern Himalayas, admittedly an easier task, in 1774. But what he did achieve was a partial solution to the long standing puzzle of the origins of the three principal Indian river systems. He showed that the Ganges did not, as Hindu tradition and current maps had it, rise from the sacred lakes, but that the Sutlej did. And further north he correctly identified a young river flowing north-west towards Ladakh as the main branch of the Indus.

But what had the distinguished vet been doing, bargaining for goats and tracking down rivers on the Tibetan plateau? He had been absent without leave from his stud and beyond the Company's territories without its knowledge. His explanation, looking for horses, was hardly convincing. Even given the then state of geographical knowledge, striking north into Tibet can scarcely have been regarded as the shortest way of reaching Balkh and Bukhara. In Tibet the only horses he was likely to see were wild *kiangs*, impossible to catch and useless for breeding purposes.

The goats made more sense. He had actually brought back fifty, four of which eventually reached Scotland where they failed to prosper. But if shawl wool could not be produced in Britain, he argued, it could at least be tried on the Indian side of the mountains and the processing and weaving of it established on British territory. This was, after all, by far the largest and most profitable commodity in Himalayan trade. The East India Company was a trading venture and it was on schemes like this that it had grown. The whole justification for Bogle's 1774 mission to Tibet had been the purchase of these goats. Moorcroft had in fact succeeded where he had failed. There might be those who disapproved of reports, which should have dealt with the progress of the stud, being devoted to goats' wool but many sympathised with the venture. And it was to provide a channel for tapping the wool trade that, after the Gurkha war of 1814, much of the hill territory south of the Sutlej had been annexed. Weaving establishments were encouraged and at Rampur a brisk trade sprang up in the 1820s. For once Moorcroft was only slightly ahead of the times.

The shawl industry remained one of his consuming interests. It epitomised his passions for trade with Central Asia and home industries for the people of both India and Britain. To the commercial drive of an eighteenth century Englishman Moorcroft united a genuine altruism. A native of Lancashire, he fully understood the plight of the working classes and, like many of his contemporaries, believed it could best be alleviated by 'improvements'. For the people of India, too, he had a real respect. Before he had reached Bilaspur on his last and greatest journey, William Laidlaw, his geologist, had been dismissed for harsh treatment of the natives. Throughout his travels one is struck by the intense loyalty shown by all his followers. There is nothing to suggest that his ideas about colonial exploitation were particularly liberal; he was simply a good man.

Trade with Central Asia was the inspiration behind all Moorcroft's schemes. He saw the area between China and the Caspian as a vast political vacuum. British influence, in the form of a network of trading interests, could bring prosperity and order whilst making of the region an outer rampart in the landward defences of India. In 1812 he had sent Mir Izzet Ullah, his confidential Persian servant, to explore a rumoured route from Punjab via Kashmir and Ladakh to Yarkand and Bukhara. In a remarkably short time the Mir faithfully performed his mission and returned to Delhi in December 1813. The track was feasible for a heavily laden caravan and armed with this first-hand account of the current political and trading position, and flushed with the success of his own first attempt to penetrate north of the mountains, Moorcroft had started to lobby the Governor-General about the necessity of crossing the Western Himalayas.

In 1813 he recommended that Mir Izzet Ullah and Hyder Young Hearsey be despatched to Kashmir and Ladakh to explore the roads beyond the mountains and gauge their feasibility as invasion routes—precisely the mission that Sir Francis Younghusband was to get all of seventy-five years later. Moorcroft argued that rumours of a French or Russian embassy in Bukhara presaged hostile activity north of the mountains and constituted a threat to British India. It was not so far fetched. The similar pretext for Elphinstone's mission to Kabul was probably in Moorcroft's mind. But the Governor-General did not concur with his Superintendent of Stud's reading of the political situation. In 1814 Moorcroft again applied, this time on his own behalf, for an expedition to Bukhara for breeding stock. It was withheld; the complications arising out of a brush with the Nepalese on the first journey had not yet been ironed out.

As soon as the Gurkha war was over the irrepressible vet was again pining for Central Asia. He discussed his plans with Hearsey. The idea was to make a second journey into Tibet by the Niti pass, proceed to Lhasa and then back to Ladakh, across the mountains to Yarkand and west along the ancient route followed by the Polos to Badakshan, Balkh and Bukhara. Even for Hearsey this was a bit much. It was too difficult, too dangerous and would never be sanctioned by the government. Hearsey had a better idea. It entailed going by sea to Bushire on the Persian Gulf and thence, with the permission of the Persians, via Teheran to Bukhara and Yarkand. But to what Hearsey called 'this safe, easy and extensive plan' Moorcroft could not agree. He held out for the Himalayan route and resigned himself to doing without his friend.

An interesting point about Hearsey's plan was that they were not all to return. George Trebeck, Moorcroft's companion when he finally crossed the Sutlej at Bilaspur, was to stay at Teheran as commercial agent. Hearsey himself was to act as agent at Bukhara and, whilst there, to help the Amir organise his forces against an anticipated attack by the Russians. This was forcing the pace of history, but it gives an indication of how both men were thinking beyond purely exploratory ventures. Precisely what lay behind these schemes will never be known, but that it amounted to more than just geographical discovery and commercial exploitation seems certain. The conduct of the journey was to give still further evidence of concealed and, to some, sinister designs.

In 1819 Moorcroft at last won permission to start for Central Asia in search of stallions. The Company had apparently run out of objections. It had certainly not had a change of heart about its Superintendent of Stud nor a change of policy towards Central Asia. Its ambivalence about the whole scheme was ample evidence. Government involvement was to be a feature of exploration in the Himalayas yet only rarely was an interest in the area publicly avowed. Moorcroft's expedition was typical. Although he carried certificates under the Governor-General's seal and was provided with specialists and an armed guard from the Company's ranks, he was refused any official status. No letters of introduction were sent, not even to the most eminent potentates like the Amir of Bukhara, and Moorcroft was expected to finance the whole operation out of his own pocket. One can understand the Company's chariness about supporting an expedition like this but, having decided to do so, the refusal to accredit it was inviting trouble, particularly with a man like the erratic Moorcroft in command. He was soon signing himself 'Superintendent of the Honourable Company's Stud on deputation to Chinese and Oosbuk Toorkistan' or '. . . on deputation to the North Western parts of Asia', convenient titles which could be made to mean as much as he liked.

His party had left from Bareilly, one hundred and fifty miles east of Delhi, in October of the same year. Lhasa had been dropped from the itinerary and their first destination was Leh in Ladakh, a corner of Western Tibet from where, according to Mir Izzet Ullah, it should be possible to make contact with the Chinese rulers of Eastern Turkestan and plan their itinerary in Central Asia. So far as they knew no European, with the exception of the Jesuit Desideri, had ever visited Ladakh. The original plan had been to tackle the Himalayas at the Niti pass, as in 1812, and to cut across the Tibetan plateau to Ladakh. But they had reached the pass too late; it was already under snow. The untried Kulu route promised to be harder in its final stages but the first part from Garhwal to Bilaspur had involved no more than tacking through the foothills. It had been accomplished without mishap during the winter months of 1819–20.

Safely across the Sutlej Moorcroft purchased a few buffalo skins and struck north towards Kulu. A day's journey took him into the small state of Sukhet. Here the altitude is only 4,000 feet and the vegetation still typically Indian. The trees, acacia, pipal and mimosa, are the wilting varieties of the plains. Bananas and sugar cane do well and the houses, though whitewashed, are mainly of mud. The difference lies in the view. Suddenly the mountains, just a white-tipped haze from the plains of the Punjab, are upon one. To the east there are glimpses of the main Himalayan chain whilst due north, and far more impressive, rises the snowy wall of the Dhaola Dhar. This is a spur of the Pir Panjal, the outermost of the West Himalayan ranges and the first which the expedition would have to cross. In March the Dhaola Dhar is at its best. The snow lies low, not on a broken succession of peaks, but on a massive ridge which, unobscured by the dun coloured foothills, rises sheer from the plains for some 13,000 feet. It was the first rampart of Moorcroft's promised land. A week's hard travelling would see them among the mountains.

There was just one possible hitch. Between the possessions of the East India Company and the Himalayan countries (Moorcroft refers to them indiscriminately as Tibet, Tartary or Ladakh) there lay a belt of small hill states. Each had its own Raja and, until recently, had enjoyed a certain independence. But since the beginning of the century they had been first terrorised by the Gurkhas and then gradually made tributary to the rising Sikh power under Ranjit Singh, Maharaja of the Punjab. By 1820 he held the key to the whole of the Western Himalayas.

Moorcroft may have hoped that by a quick dash through Mandi and Kulu he could run the Sikh gauntlet. Certainly in Sukhet, the first of these states, there was no trouble. The villagers fled at their approach but soon flocked back to stare at the 'feringhis' and their camp style; they had never seen a European before. The night was disturbed by nothing worse than the howling of hyenas and morning saw them pressing on over the rolling countryside to Mandi. Somewhere near the new town of Sundernagar they were stopped by a tatterdemalion rabble armed with swords, matchlocks and bows and arrows. It was the Raja of Mandi's army. The Raja had no objection to their crossing his territory but some Sikhs collecting revenue in the area had. They were adamant that no one could pass without reference to the Maharaja in Lahore.

Here as elsewhere the trouble was caused by the expedition being far too big. Anything like an inconspicuous dash was out of the question. Moorcroft had with him over £3,000 worth of merchandise, most of it heavy bales of cloth necessitating scores of porters and muleteers. Partly because he believed he could make a profit for his backers, partly to sustain his assumed role as a bona fide trader and partly just to help the expedition's finances—he always had too many good reasons for a disastrous idea—he had lumbered himself with the impedimenta of a full-scale caravan. This was to be a sore temptation to the tribes in the more unsettled areas whilst here it was the hordes required to guard and carry it which caused consternation. Wild rumours began to circulate that a 'Feringhi' army had crossed the Sutlej, taken Mandi and proclaimed that the hill states were under British protection.

In the hope that contact with Ranjit Singh might be turned to good account, Moorcroft himself, along with the invaluable Mir Izzet Ullah, had set off for Lahore. But the rumours overtook him as he made his way across the baking plains and at the unlovely town of Hoshiarpur he was arrested. For the whole of April he sweated it out in a doss-house for beggars beside the town's cess pit, whilst the Mir sped on ahead to arrange his release. At last, in early May, he was sent under guard to Lahore. The heat was now so great that they could travel only at night. Much of the way the ageing vet was carried in a *jampan*, a sort of hammock slung from a pole. In

his weakened state he turned vegetarian and took dinner only three
times a week.

One sympathises with Moorcroft's discomforts and irritations as
he grapples with the hot weather and the whims of a hypochondriac
Maharaja, but at least he was doing something. Most of the party
were just stuck in Mandi. This is a busy little town on the Beas
river; it could be very pleasant. The riverside is strewn with small
stone temples, there is a good bazaar and a fine old hill palace, like a
colossal timbered barn, overlooking it. But the visitor's recollection
is more often one of heat and flies and frustration. From all directions
arid grey hillsides slide steeply to the river. The town nestles between
them sheltered from the erring breeze and deprived of any view. At
6,000 feet it might be charming, but at 2,000 feet, for most of the year
and especially during the hot season, it is hell. The air is so still that
smoke and smells just hang there; the flies are so lethargic they have
to be pushed away.

In charge of the main party at Mandi was George Trebeck, the
son of a solicitor friend of Moorcroft's and his only European companion. (Guthrie, the party's doctor who had been lent along with
Laidlaw, the intolerant geologist, by the Bengal government, was
an Anglo-Indian.) Moorcroft always refers to Trebeck as 'my young
friend' while the latter speaks of his leader as 'Mr. M.'. There was
thirty years between them. One thinks of Trebeck as a disciple rather
than a companion, a role which fits well with Moorcroft as the
prophet. His letters from Ladakh will sound much like the voice
of one crying in the wilderness.

For the three hottest months of the year Trebeck sat it out at
Mandi. It must have been a trying time. The Sikhs would have been
only too pleased to see him slip back across the Sutlej. But even with
the prospect of having to tackle the journey through Kulu at the
height of the monsoon he did not waver. In June, just as the rains
were starting, came word from Moorcroft to proceed. The tents were
struck and the porters loaded. Then once again the Sikh commandant closed in on them. The camp was invested more closely than
ever and it was not till a month later that they were finally free to
move out. In torrential rain they reached Kulu on July 19th.

Moorcroft joined them there two weeks later. He had been recalled
to Lahore a second time and after further delays and prevarication
had slipped away to the court of Sanser Chand, Raja of Kangra.
Kangra had once been the largest of the hill states and Moorcroft

listened with sympathy to the old Raja's tales of the Sikh perfidy which had destroyed it. His own growing dislike of the new power in the Punjab turned to an indefatigable hatred which was to colour all his subsequent dealings in the Himalayas.

Between the two men there grew up a close friendship and, when Moorcroft cured the Raja's brother of apoplexy, Sanser Chand's generosity knew no bounds. His guest was to select any tract of land in the country that might appeal to him, any horse from the royal stables and any damsel from the Raja's dancing troupes. The offer of land was one he preferred to leave until his return, though the idea of retiring to the foothills already attracted him. In the Raja's stables there was probably little to excite the Honourable Company's Superintendent of Stud but the *nach* girls, the famed dancing girls and courtesans of northern India, were not beyond his interest even at fifty-five. He always maintained that their acquaintance was worth cultivating since they were the best informed and most communicative natives of any the traveller might meet. He dallied at Kangra for a full six weeks and then sent back from Kulu 'a fine gold-embroidered muslin for Jumalo, his favourite'.

From Kangra he had crossed to Baijnath and then via the Dulchi pass descended on the Kulu valley. His spirits were high 'having now before me no further obstacles to my penetrating to Tibet [Ladakh] than the natural difficulties of the country and the weather'. Torrential rain and mud made the mountain tracks treacherous but Sanser Chand had helped out with a hundred porters and when they emerged above the clouds on the Dulchi, Moorcroft was as near poetic rapture as his scientific training would allow.

> . . . vast slopes of grass declined from the summits of the mountains in a uniform direction but separated by clumps of cedar, cypress and fir; the ground was literally enamelled with asters, anemones and wild strawberries. In some places the tops of the hills near at hand were clearly defined against a rich blue sky whilst in others they were lost amidst a mass of white clouds. Some of them presented gentle aclivities covered with verdure, whilst others offered bare precipitous cliffs, over which the water was rushing in noisy cascades. In the distance right before us rose the snowy peaks, as if to bar our further progress. Vast flocks of white goats were browsing on the lower hills and every patch of tableland presented a village and cultivated fields; glittering rivulets were

meandering through the valleys and a black forest of pines frowned beneath our feet.

They were soon back in the clouds and it poured throughout the four day journey through the Kulu valley. They saw little of one of the most scenic valleys in the whole of the Himalayas. The challenging peaks above the steep wooded hillsides were completely obscured, though the world down below evidenced significant changes. The people were now dressed in tweeds instead of cottons and Moorcroft marvelled at the tartan-like plaids and homespuns. The women were small and pretty; their houses substantial affairs of stone and timber with deep verandahs hung with trailing vines. Orchards were plentiful. The air was full of the noise of rushing mountain torrents and the scent of pine and cedar. Mostly they kept to the left bank of the Beas crossing to the right near the modern tourist resort of Manali and recrossing to sample the hot springs of Vashisht. Had there now been a break in the clouds they would have seen not a patch of blue sky but a craggy chunk of the Pir Panjal. They were right underneath it.

The Kulu valley ends and the Beas river begins on the Rohtang pass, a saddle at the eastern extremity of the Pir Panjal. As Himalayan passes go it is not high, a mere 13,000 feet, but for all that a killer. Sudden icy blasts still take their toll of those who cross too late in the year. Colossal snow drifts and landslides still defy all efforts to keep open the newly built road for more than six months in twelve. In the Western Himalayas the most southerly range attracts the highest precipitation while the higher peaks and plateaux behind remain comparatively dry and snowfree, though invariably colder. Thus the Pir Panjal at the Rohtang pass forms the dividing line between the lush Indian hill country and a distinctly harsher world. The magnificent forests of sycamore, oak and cedar on the southern slopes are represented to the north by a rare and stunted cypress clawing at the rocks or a few heavily pollarded willows adorning the villages. The moist green pastures and emerald rice fields of Kulu are replaced by bare boulder-strewn scarps all orange and brown. There is no handy perch for the ubiquitous Indian crow; instead there is the lighter and shyer chough, cavorting in air that looks too thin for flight. The dazzling sun strikes down, as much sharp as hot, and burns like acid. It must be one of the most dramatic physical changes on the face of the earth.

The people too are quite different. The Kuluis are Hindus of Aryan race. Lahul, the land immediately beyond the pass, has a predominantly Buddhist population of distant Mongol descent. Its square, mud-faced houses stacked one above the other up the steep hillsides are the same as those of Ladakh which would remind Moorcroft of card houses. The fluttering of prayer flags and the muttering of lamas, assisted by the ceaseless rustling of the leaves on the lofty poplars, evidence the busy piety of the Buddha's followers. In all but name Tibet starts at the Rohtang.

From this pass to Leh the party was on the move for two months. It was one of their longest spells of continuous progress. They crossed the Rohtang, pressed on through the Lahul valleys, struggled up the Bara Lacha pass over the Great Himalaya and then across the western extremity of the Tibetan plateau to the Indus and Leh. Beyond Lahul the country was bleak and uninhabited. It was bitterly cold, but the going was comparatively easy and the expedition at last settled down into a regular routine.

There was no excuse for getting lost. To the people of Lahul and Ladakh this was a well-known route and they supplied guides, porters and mules. But with most of the men on foot progress was slow and, in a land where the clarity of the atmosphere is such that one can start for a nearby ridge at dawn and be no noticeably closer to it by dusk, it must have seemed even slower. Moving in single file because of the narrowness of the track, the column stretched for at least a mile. In front would be the strings of mules, skinny beasts with many a bald scar and daily worsening sores where their heavy loads rubbed; but for all that strong, sure-footed and well acclimatised. The soft ringing of their thick brass bells, added to the tap of their shoes against the rock or the tinkle of falling stones when a hoof slipped, only intensified the oppressive silence of the mountains.

Behind stretched the straggling line of porters. In the plains heavy loads are carried on the head but in the mountains men and women will carry everything, from the lightest handful of fodder to the heaviest of tree-trunks, on their backs. Small articles are jammed in a conical basket slung from the shoulders; a sack of grain or a tree-trunk is ingeniously tied round the waist and shoulders with a single rope. The weight of each load may be as much as 80 lbs. Bent beneath it, neck sinews prominent and head straining upwards, the wiry little porter looks much like a tortoise. As the altitude increases so does the frequency of his halts. Any rock of a convenient height to support his

Rope bridge in the Pir Panjal from G. T. Vigne's *Travels in Kashmir, Ladak and Iskardo.*

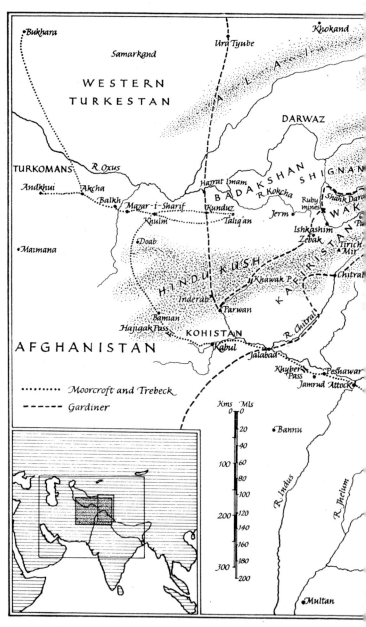

The Western Himalayas showing all the principal mountain system
Gardiner. (To cover chapter

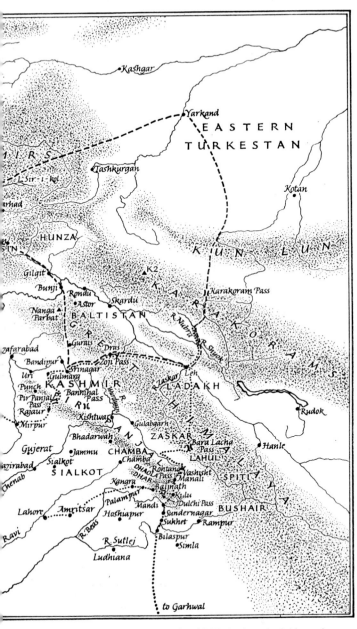

…d illustrating the routes of Moorcroft and Trebeck and of Alexander
…9 excluding part of chapter 5)

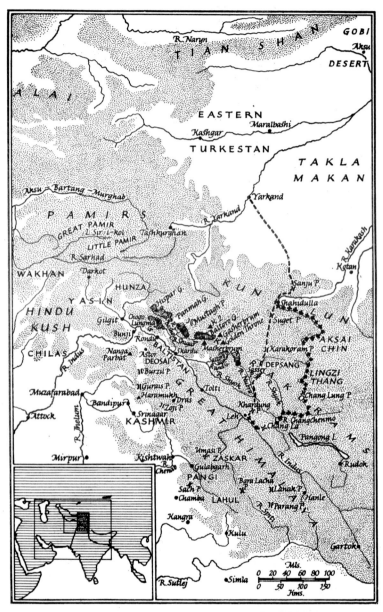

Baltistan and the Road to Yarkand showing the Karakoram and Changchenmo routes and the main Karakoram peaks and glaciers. (To cover chapters 10–14 and part of chapter 5)

load is rubbed smooth with use. He talks little but a soft muttering seems to accompany each long-drawn breath—a repetitive prayer perhaps or some tuneless refrain.

A more purposeful mumbling came from the two Indian *pandits* specially employed by Moorcroft to measure the route. As on the 1812 trip they were expected to take a regulation four-foot stride. Such a giant pace had provoked a good deal of ridicule from armchair geographers and Moorcroft was at pains to point out that by a stride he meant a pace forward with each foot. Most of the geographical work was left to Trebeck who had had some training in surveying. He, like Moorcroft, was mounted and would spend the day's march dashing from one vantage point to the next taking compass bearings and cross bearings and checking on the progress of the *pandits*. For measuring latitudes he had a sextant but longitudes were more difficult because of 'the northing of the sun'. His only other instruments were a thermometer and a barometer from which, by measuring the temperature of boiling water, he was able to ascertain the height of the passes. The highest, the Bara Lacha, he made 16,500 feet above sea level. In the howling gale which here, as so often, accompanied the crossing of the pass, this method of measuring altitudes was a most unpleasant business, but it remained the usual one for the next fifty years.

Meanwhile Moorcroft spent his time preparing a dossier on the country. As a traveller he is the absolute antithesis of the modern tourist. The idea of passively marvelling at the wonders of nature is meaningless to him. So much so that he fails to record the wild spendour of the Lahul valleys or the extraordinary clarity of the air. He is much more interested in coming to grips with life in these surroundings. No village headman, no farmer or traveller is allowed to pass without Moorcroft pouncing on him with a volley of questions. What is he carrying, what is he growing, for how much will it sell, for what is it used? In the absence of more specific instructions it seems that Moorcroft must have recalled those given by Warren Hastings to George Bogle on his attempt to reach Lhasa in 1774. Hastings wanted to know everything; no detail of animal, vegetable or mineral life was to be ignored. Bogle did his best but Moorcroft could have done better. His curiosity is unbounded and his notes read more like a gazetteer than a traveller's journal. In a day's march he found more to record than another man would of the whole journey. His thirsty pen scuds across the foolscap; the sloping hand leans ever lower under the pent-up deluge of ideas bearing down upon it. It plunges over an invisible lip and on down the side of the sheet in a cataract of words that won't be interrupted, even for a new line.

Yet in all these outpourings one looks in vain for the personal touches that should enliven a travelogue. He is in fact a wretched

Rope bridge on the Chandra

travel writer. Far from magnifying the difficulties for the benefit of the reader he is at pains to gloss over them. When two days are spent trying to get horses across the raging Chandra river one wonders whether they were not within a whisker of having to turn back to Kulu. And when he lets slip that on the Bara Lacha his ruddy face was raw and torn and that he suffered from a headache, one is tempted to diagnose one of the first recorded cases of high-altitude pneumonia. It was partly reticence, but also Moorcroft wants us to think that his new route is ideal for the busy trade he expects to open up with Central Asia. There was too much at stake for light asides and awkward scrapes. In what was to be one of the greatest scientific and commercial undertakings of the century a certain dignity was desirable.

With less by way of excitements than would attend any other recorded journey along the same route he reached Leh, halfway through the mountain barrier, in September 1820.

2. The Unfortunate Moorcroft

Although it had taken a whole year Moorcroft's journey as far as Ladakh had been a success. As yet there was no Royal Geographical Society to acknowledge it; even the concept of the explorer was not current. Yet the achievements spoke for themselves. Penetrating further into the Western Himalayas than any before him he had discovered and pioneered a possible trade route to Leh. He had strengthened his claim as the greatest authority on the Indus by adding to his discoveries of 1812 the sources of the Beas, Ravi and Chenab tributaries plus another long stretch of the parent river. And, as the first Englishman to visit Kulu and Ladakh, he had carried the British name as far towards Central Asia as anyone. When the information started to filter through to the government in Calcutta they must have congratulated themselves. Already the results justified their minimal commitment and, better still, there had been no political repercussions.

Now things were to change. Moorcroft spent three years in Ladakh and in Kashmir. The delay was in itself enough to try the patience of his employers. It perhaps warranted the dishonouring of his money drafts on Delhi which caused him 'anxiety oppressive beyond expression'. But this was nothing to what was to follow. The man who emerged from Kashmir in 1823 was not just harrassed but bitterly disillusioned. If he continued to wax sanguine about the outcome of the expedition it was from defiance as much as enthusiasm.

What had gone wrong? Practically everything. In the first place his original plan of proceeding to Yarkand had to be abandoned. He successfully concluded a trading concession with the Ladakh authorities, but it was of little value if a similar arrangement could not be made with the Chinese. Ladakh after all was a poor and sparsely populated place; its commercial importance was primarily as an entrepôt. The real markets lay beyond its northern mountains in the populous and wealthy cities of Turkestan. Yarkand in Eastern Turkestan, a land then—and now as Sinkiang—under Chinese rule, was the nearest of these. Mir Izzet Ullah was sent there to prepare the way and it seemed for a time that Moorcroft's chances of getting permission to follow were good. Through the summer and winter of 1821 he waited. A stream of messages was sent north and every trader who found his way south to Leh was collared in the long open bazaar and pumped for news. But the Chinese were traditionally suspicious of all visitors. There was no precedent; reference must needs be made to Kashgar, to Pekin. Equally the Kashmiri merchants in Leh, who held a monopoly of the existing trade, were averse to a British intrusion. An attempt on Trebeck's life was probably their doing, and it was their report to the Chinese claiming that the British party had political ambitions which finally clinched the matter.

If permission had been granted it must remain doubtful whether Moorcroft could have successfully tackled the five terrible passes that lay ahead. Though the distance was only four hundred miles he seems to have had little idea of the hazards involved or of how much more difficult was the crossing of the Karakorams and Kun Lun than that of the Pir Panjal and the Great Himalaya. Almost fifty years would elapse before an Englishman finally made it to Yarkand.

Moorcroft was very hard on the Kashmiris. He called them 'the most profligate race on the face of the earth'. But, in fairness, they had a point. For someone who professed to be a simple trader his behaviour in Ladakh was very strange. First there was the affair of Aga Mehdi, a Kashmiri Jew by birth, who was ostensibly trading between Ladakh and Russia but who, it was thought, acted as agent both commercial and political for the Tsar. This was so sinister that Moorcroft became convinced that the Russians, whose frontier at the time was all of 2,000 miles away beyond the Aral sea, had designs on British India. Did their agent not carry gifts and cash even for Ranjit Singh, the most formidable power on the Indian frontier? And were there not four hundred cossacks waiting at the frontiers of Turkestan to escort Aga Mehdi and envoys from Ladakh and Punjab back to St. Petersberg?

Letters from the agent, then on his way south, arrived in Ladakh. Unfortunately they were written in Russian and Nogai Tatar, languages unknown to anyone in Leh. Trebeck set to to make a tracing of the unfamiliar scripts and sent it off to Calcutta. Moorcroft began to see Aga Mehdi as his opposite number, 'a shrewd and able competitor', and through the long Tibetan winter of 1820-21 Russia's designs preyed on his mind. Even in his house in Leh the temperature remained for weeks below freezing. He was convinced that nothing less than the safety of India lay in his hands and, plagued by rheumatism, he prepared for the battle of wits that would come with the agent's arrival in the spring.

But Aga Mehdi never reached Ladakh. His companion and his caravan arrived in April but he himself lay dead in the mountains 'of a sudden and violent disorder'. For Moorcroft it was most disappointing. Had he only lived 'he might have produced scenes in Asia which would have astonished some of the cabinets in Europe'. Most of his papers had disappeared and Moorcroft had to be content with a letter from the Imperial court to Ranjit Singh. This, too, was in Russian, but by chance in the summer of 1822 Moorcroft met the Hungarian Csoma de Koros,* who was able to translate it. It was little more than an introduction and, though interesting evidence, did nothing to confirm Moorcroft's speculations. In spite of encouragement from the British party Aga Mehdi's companion refused to continue to Lahore and discharge the dead man's mission. He preferred to sell the rubies intended for Ranjit Singh and make off with the profits. Without the hard proof he had hoped for, Moorcroft's speculations looked like extravagant alarmism and provoked a suitably sceptical response in India.

Infinitely more serious in the eyes of his employers was the zealous but ill-advised politicking between Moorcroft and his Ladakhi hosts. At the time of his visit the political status of Ladakh was vague. Its people, language, religion and geography are Tibetan. Probably Tibet had originally been unified from Leh and there were still close ties with Lhasa involving spasmodic tribute and frequent religious contact. On the other hand it had also sought protection from the Moghuls in India and paid tribute via Kashmir to the

* The Hungarian scholar, Alexander Csoma de Koros, had armed himself with a stout walking stick and from his native Transylvania set off for China. Two years of wandering across Asia, of which no more is known than his itinerary, brought him to Kashmir. En route from there to Leh he ran into Moorcroft. The meeting changed his whole life. Moorcroft became his patron, providing money and introductions on the strength of which he devoted the rest of his life to the study of the Tibetan language and of the giant volumes locked away in the monasteries of Ladakh. After Moorcroft's death the government rather reluctantly continued his stipend. He was scarcely a financial burden; he lived off Tibetan tea, had one change of clothes and was too absorbed in his studies to notice the frost biting at his toes. He rewarded Moorcroft's kindness by eventually producing the first Tibetan dictionary and grammar; he was the leading Tibetan scholar of his day—and the only Tibetan scholar of his day. See T. Duka's *Life of Csoma de Koros*.

Emperor Aurangzeb. Moorcroft joyfully pounced on this, declaring to a correspondent, 'Ladakh is tributary to Delhi!' And Delhi in the person of the puppet Moghul was now under British protection. By the happiest of coincidences it seemed that this unknown country was already British territory.

When the Sikhs, who had taken Kashmir in 1819, demanded from Ladakh the customary tribute, Moorcroft therefore advised against it. The Ladakhis should deal only with Delhi. They were by now wholly in his confidence and he in theirs. After negotiating the commercial arrangement they were convinced that Moorcroft was an emissary of standing. What more natural than that they should take advantage of his presence to snub the Sikhs and officially proclaim their allegiance to the British? It was not the first time Moorcroft had been approached in this way; both Kangra and Kulu had also sought British protection. But it was the first time he did anything about it. With the supposed implications of Aga Mehdi's dealings in mind, he saw this chance of establishing British commercial and political authority in the Western Himalayas as heaven-sent. Moreover Leh was not only the key to the trade of Central Asia but, given access to Yarkand, also to inland China. Here was the long-sought back door into that merchant's paradise which could change the whole pattern of Sino-British relations. In the longest of longhands Moorcroft covered reams with arguments. 'Maharajah Runjeet Singh seeks the allegiance of Ladakh by invitation, intrigue and menace. Russia invites the allegiance of Ladakh by promises of commercial advantages, of titles, of Embassies of distinguished honour. To the Honourable Company Ladakh, unsolicited, tenders voluntary allegiance!!' It was irresistible. How could he do other than forward Ladakh's formal request?—he had probably drafted it too. At the same time, and this was his big mistake, he wrote to Ranjit Singh telling him of Ladakh's status vis-à-vis Delhi and advising him not to meddle further in its affairs until he heard from the British government. It was not a threat; Moorcroft was just playing for time. But it could have been read that way.

For the British authorities this was too much. The Resident in Delhi was so amazed he could not trust his own translation of the note sent by Moorcroft to Ranjit Singh. Their Superintendent of Stud had no authority whatever to meddle in the Company's external relations. Ladakh played no part in their schemes; it was too remote and too poor. Ranjit Singh, on the other hand, was a powerful and sensitive ally. It was understood that, in return for not intriguing south of the Sutlej, he should have a free hand in the west and north. An apology was swiftly forwarded to Lahore. Moorcroft was severely reprimanded and his conduct disowned. A few months later his salary was stopped and he was recalled to Calcutta. Official notification of this came in 1824, too late to catch the party before they had headed off to Afghanistan, but rumours of it reached them in Kashmir. 'It is a strange tale that Moorcroft has been recalled by the Court of Directors . . .' wrote the loyal Trebeck. 'If it were a fact they know him not who suppose he would obey the summons to return.'

Whatever the government's doubts, Moorcroft had none. It is clear that throughout he acted in the best possible faith. His reasons for accepting Ladakh's offer, though too many and too elaborate, were far-sighted rather than far fetched. But there was another reason, unexpressed. He genuinely loved the Ladakhis. They became his adopted people, and it is not without pride that he tells of a move to have him made ruler of the country. During their two years there he and Trebeck explored it perhaps more thoroughly than any subsequent visitor. They saw the comparatively verdant valley of Nubra towards the Karakoram pass, and the desert wilderness of Changthang and the Pangong Lake on the Tibetan side. They explored the Dras valley on the Kashmir frontier and the Spiti valley to where it marched with the British protected state of Bushair. And they grew to like this skeletal land where the lowest point of the warmest valley is all of 9,000 feet above sea-level.

Most of it looks like a country without a countryside, a land from which all soil and vegetation has been torn away, leaving just the bare bones of jagged rock and beetling cliff. The valleys are not the deep sheltered clefts of the Pir Panjal but open troughs along whose flat bottoms the dust-laden wind has free play. The villages are perched high up on the sides on shelves of alluvium and separated from one another by steep slopes of what look like railway clinkers. Yet Ladakh's first visitors far preferred it to Kashmir—'an odd choice' wrote Trebeck, 'but there is something fine in the dash of a torrent and the wildness of a mountain desert'.

Everywhere they were met with kindness. The people were the antithesis of their surroundings, warm, gentle and welcoming. Moorcroft's medical work brought a flood of patients who turned his house into a market place with their offerings of vegetables and livestock. In gratitude for a minor operation, a patient's old mother would come a week's march to present him with a sheep. They were as well received in the lama's cliff-top monastery as in the shepherd's stone shelter or the tall palace in Leh. There was, and still is, something wholly idyllic about this smiling and not uncivilised people in their remote and harsh surroundings. Not surprisingly Moorcroft wished to save them from the heavy yoke of the Sikhs.

Practical as ever, what impressed him most was their husbandry. Page after page of his journal and letters are devoted to their rhubarb, their barley, their minuscule sheep, their cattle fodder, bee-keeping, manuring and so on. It was the same story in Kashmir and most of the way to Bukhara. Trebeck calculated that if all Moorcroft's schemes to introduce new systems of agriculture, new crops, cattle and industries were adopted, the country would be 'gainers by a million a year'. Poring over those ill-written journals today, all of 150 years later, the reader might wonder whether any agronomist has yet sifted through them. Trebeck was exaggerating but Moorcroft's observations were seldom as crazy as they might look. He was advocating the export of Indian tea to Central Asia before it had been successfully grown in India. Yet fifty years later the man who opened trade with Yarkand would be an ambitious tea planter with surplus production.

Moorcroft was thinking increasingly of the future and one suspects that many of these observations were recorded as personal *aides memoires*. Both in Garhwal and Kangra he had looked for a suitable tract of land for his retirement. In Leh he drew up a detailed list of furniture and fittings for the house-to-be, making sure that there was nothing that could not be carried by porters; he obviously had an isolated site in mind. It was not so much a building plot that he wanted as a small territory. Surely, he asked, there must be a tract of undeveloped hill country somewhere that the government would cede to him in expectation of an increased revenue once his improvements got going? He was prepared to guarantee that the foothills of northern India could be as prosperous and productive as anywhere in the country.

* * *

In September 1823, four years after setting out from Bareilly and three after reaching Ladakh, Moorcroft resumed his journey. During his long stay in the mountains he had made innumerable excursions but had progressed only as far as Kashmir. He is not a good guide to the famous valley. The ten months he spent there were unusually wet, the Kashmiris, whom he cordially detested anyway, were experiencing a famine, and the oppression of their Sikh overlords again invited his contempt. The exploratory forays continued but most of his time was devoted to one of the most detailed examinations of the shawl trade ever attempted. His notes are still the most authoritative on the various production processes. Victor Jacquemont, the next European to enter Kashmir, claims that 'his chief occupation was making love and, if his friends are surprised that his travels were so unproductive, they may ascribe it to this cause'. This was before the publication of Moorcroft's travels. As will be seen, coming from Jacquemont it is a prize piece of hypocrisy. Certainly the old traveller had his troupe of *nach* girls but they scarcely seem to have interrupted the customary deluge of observations. Besides, it was to be his last chance to relax, and where better than in the balmy atmosphere of Kashmir? It was about this very time that Moore was penning his celebrated lines:

> If a woman can make the worst wilderness dear,
> Think, think what a heav'n she must make of Cashmere.

Perhaps to combat the effect of the valley's lassitude he also formed a private army. Trebeck wrote off to India for drill manuals, arms were both made in Kashmir and ordered from the British political agent on the Sutlej,* and Moorcroft had his men out in the Srinagar suburbs doing manœuvres twice a week. It was a sure way of arousing the suspicions of his Sikh hosts. Dr. Guthrie and the Mir objected strongly but Moorcroft, forbidden the cherished route via Yarkand and already in bad odour with the Sikhs, no longer worried. For the final bid to reach Bukhara he had resolved to adopt an altogether firmer line. There was to be none of the docility he had shown in Mandi in 1820. When their path was barred by Punjabi miscreants just across the Indus he drew up his thirty cavalry in battle order and unwrapped a miniature cannon which had been intended as a present for the Amir of Bukhara.

Travelling by way of the Pir Panjal pass, and avoiding the embarrassment of another rencontre with Ranjit Singh at Lahore, they made

---

* The agent in question was Captain Charles Pratt Kennedy. He had just moved into his new house, high above the river Sutlej in a village called 'Semla'. It was the founding of the future summer capital of British India.

for Peshawar and halted there a modest five months. This was the beginning of Afghan territory, then in a chronically disturbed state. If the party had roused a few suspicions in the Punjab hill states, here its motives were to be invariably misconstrued. To the Afghans it was beyond comprehension that an Englishman could venture into their troubled country simply for trade and horses. Either Moorcroft must be an official emissary like Elphinstone or possibly a soldier of fortune, or else he was a spy.

At Peshawar all went well. Moorcroft's hosts shared his dislike of the Sikhs. They also saw nothing but advantage in having a British representative in their ranks on an imminent bid for Kabul. To cement matters, Moorcroft was offered the governorship of the city and invited to tender its allegiance to the Honourable Company. This time, not surprisingly, he fought shy, but Trebeck seems to have been within an ace of assuming the governorship. Remembering how Hearsey's plan had allowed for his remaining at Teheran as commercial agent, it is possible that Moorcroft, who did not oppose the plan, envisaged a similar role for him at Peshawar. However, the offer fell through and, with their neutrality compromised, they continued to Kabul in the van of the Peshawar forces. Again they were lucky. Pitched battles were raging in the streets of the city but, for a substantial 'loan', their Peshawar allies remained firm and saw them safely on their way to the Hindu Kush. They crossed by the Hajigak pass and Bamian reaching Khulm in early September 1824. Here their luck finally ran out.

For nearly six months they were held by the chief of Kunduz, Murad Beg, compared to whose treatment Ranjit Singh's had been positively angelic. Throughout they were in daily fear for the future of the expedition, their merchandise and their lives. Reading Moorcroft's papers one gets, for the first time, a sense of impending disaster. Events, suddenly, are being telescoped. Interludes on the agriculture and curiosities of the region become scarce, while a host of new names and complex intrigues take their place. Since Peshawar the whole pace of the expedition has speeded up. Meals of a simple Indian character are taken squatting on the ground; table and chairs are no longer unpacked and Moorcroft's bed has been lost between Leh and Srinagar. There is a growing impression of disintegration. Trebeck is unwell, Guthrie becomes separated from the party and Mir Izzet Ullah, the most invaluable of all, despairs of success and, sick and dying, is sent back to Peshawar. Moorcroft, too, is a changed man. Cornered, he acts and writes with impulsive and uncharacteristic vehemence. Every rebuff from Murad Beg is a threat not just to his safety but to his cherished plans of trade with the region. In his anxiety he even loses track of the dates.

Murad Beg's object was by fair means or foul to mulct the party of everything they possessed. He knew they had valuable merchandise, a considerable sum of money and probably some political standing. The idea was to frighten them into parting with the first two and, before finally deciding how to dispose of his captives, into revealing the precise nature of the third. From Khulm to Kunduz is a seventy-mile ride over low hills and desert. Three times Moorcroft made the journey there and back. After presents, bribes and an arrangement about customs dues he would return to Khulm only to be again prevented from leaving. Gradually it became clear that they were prisoners. Moorcroft no longer minced his words. To his face he accused the Uzbek chief of hypocrisy and he threw down the letter promising safe conduct. Murad Beg exploded with rage and muttering something about infidels left the court. Only 50,000 rupees was now to save them from 'a taste of the summer at Kunduz'.

Back in Khulm Trebeck was all for fighting it out. Though hopelessly outnumbered, he reckoned that their drill in Kashmir plus surprise tactics would see them through. Whatever the outcome it would be better than this slow strangulation. But Moorcroft had another idea. He would seek sanctuary and help from the one reputedly honest man in the country, the Pirzada of Taliq-an, whom even Murad Beg respected. It involved escaping from Khulm, riding undetected for one hundred and fifty miles and chancing everything on the generosity of a man he had never met. But the risk was just his own and in the end it paid off. Heavily disguised and riding continuously for two days and two nights through heavy mud, he made it to Taliq'an. The Pirzada gave him sanctuary, listened to Moorcroft's tale of woe and, after an open disputation in which his accusers maintained he was commander-in-chief of the British forces in India, decided in his favour. He interceded with the chief and on February 3rd 1825 Moorcroft, poorer by only 2,000 rupees, left Kunduz territory. They had escaped. But the awkward fact remained that they had to return. Murad Beg's rule stretched along the south bank of the Oxus from Badakshan to Balkh. It would be almost impossible to by-pass it.

Meanwhile there was Bukhara, Bukhara the Noble, perhaps the

oldest, most revered and least known city in Central Asia. Even Marco Polo had not been there. Surprisingly Moorcroft and Trebeck were not the first Englishmen to visit it—Elizabeth I had sent an envoy—but they were the first for two hundred years. In all their schemes and wanderings it had been the ultimate goal, the hub of the land trade of Asia and the market for the horses which, if all else failed, would ensure the expedition's success. Trebeck compared their arrival to the Crusaders' first sight of Jerusalem. 'After a long and laborious journey of more than five years we had a right to hail the domes and minarets with as much pleasure as Geoffrey de Bouillon.'

Unlike later travellers they were well received. Permission was given to ride through the city, an unheard-of privilege for a non-Mohammedan, and the Amir gave them freedom to trade. They disposed of most of their merchandise and procured a hundred horses. This was not as many as they had hoped, but their performance fully lived up to expectations. The only cloud on the horizon was the prospect of another brush with Murad Beg and, unwisely, they gave vent to their feelings about him to the Bukharan chief minister. In late June they left the city heading back to India.

By November it was all over. Moorcroft, Trebeck and Guthrie were dead. So too, in Kabul, was Mir Izzet Ullah. Their horses, property and servants were dispersed through Afghanistan. Rumours of the disaster started to reach India at the end of the year, but it was not till the following May that a letter arrived from Trebeck confirming Moorcroft's death and not till 1828 that the sole survivor to reach India, Hearsey's man Ghulam Hyder Khan, found his way to British territory.

The mystery of just what happened is now almost certain to remain unsolved. For reasons that it is difficult to understand, no attempt was made at the time to establish the real facts. There seems, on the contrary, to have been a conspiracy of silence. Ghulam Hyder Khan's account of the expedition was serialised in the *Asiatic Journal* but not till ten years later and even then the final section, dealing with events after they left Kabul, was suppressed. A report from that city about the death of Trebeck in which the writer pleaded that 'a good, honest and trustworthy man' be sent immediately to retrieve the horses was ignored. The Royal Geographical Society in London, founded in 1831, published some of Moorcroft's notes from Leh and Kashmir in the first issues of their journal. But at the time they were still uncertain whether he had ever reached

Bukhara, let alone the circumstances of his death. When in 1840 Alexander Burnes discovered some revealing private letters from Trebeck, including one from Bukhara, he forwarded them to the Society for publication. Inexplicably they were neglected.

It was not till 1841 that Moorcroft's *Travels* were finally published. By then others had covered most of the ground so that they contained little that was original. Equally what was contentious in them was no longer so. This presumably was just what was intended. But who was to blame? Was it the Company who made good Moorcroft's arrears of salary in order to get possession of the papers and who were as anxious for his death to be overlooked as they had been embarrassed by his political activities? Or was it Moorcroft's successors, anxious to ensure maximum acclaim for their own achievements? Sir Alexander Burnes brought Moorcroft's papers back from India in 1832 and did much for his reputation, but only after the publication of his own *Travels to Bokhara*. In 1835, when the editor of Moorcroft's journals, H. H. Wilson, finally had the Travels ready for publication, William Fraser, a British official in Delhi, was murdered. Amongst his papers were found seven volumes of Moorcroft's journal dealing with the final stages of the expedition. How Fraser had come by these and whether he was holding them for the government or witholding them from the government is unknown. We do, however, know that he was the closest of friends with Victor Jacquemont, Moorcroft's successor in Kashmir, and had himself planned to cross the Sutlej.

It was another three years before Wilson had incorporated the new material and another two before it was published. Even then Wilson mysteriously stopped his account with Moorcroft's arrival at Bukhara. He completely ignored the last volume, 'Return from Bukhara', which had definitely been in Fraser's possession and was therefore known to Wilson.

'Return from Bukhara' shows the party setting off in a frame of mind that is still sanguine—Moorcroft's favourite adjective. In the long dissertations on the guinea worm, the raisins of Bukhara, cockfighting and a novel way of cooking peaches we recognise vintage Moorcroft with its heavy sediment of adventitious matter. He is feeling the heat, his face is redder than ever and his lips badly blistered. But it is no worse than on the Bara Lacha. For a man of sixty who has been travelling for five years he sounds as fit and alert as ever.

They recrossed the Oxus about August 4th and at Akcha were rejoined by Askar Ali Khan, a henchman of Mir Izzet Ullah, who had been sent back to Peshawar to explore a possible return route via Badakshan and Chitral. In this he had been successful. They were thus in a position to give Kabul a wide berth, if not Kunduz. Askar Ali was now sent to feel out Murad Beg while Moorcroft, alone but for a few servants, set off across the desert, west to Maimana. In Bukhara the shortage of horses had been blamed on the disturbed state of the surrounding country. The times were too dangerous for them to be brought to market, but anyone willing to risk his neck could purchase them from their Turkoman breeders in the desert oases to the south. Moorcroft decided to do just that. He had written from Bukhara, 'Before I quit Turkestan I mean to penetrate into that tract which contains probably the best horses in Asia but with which all intercourse has been suspended during the last five years. The experiment is full of hazard but *le jeu vaut bien la chandelle.*' The candle, as one of his later admirers puts it, had not long to burn now. At Andkhui, in the very back of beyond, he was taken ill of fever and on August 27th 1825 he died. There were rumours of poison and of 'being struck in the chest by a bullet' but the body which was brought back to Trebeck would have been beyond recognition by the time it arrived.

The next to go was Guthrie. Also suffering from fever, he was carried from Akcha to Balkh on a stretcher and there he too died. Trebeck, now alone and himself sick, was a broken man. While 'Mr. M.' lived, he was prepared to risk anything. Without Moorcroft and assailed by reports of Murad Beg's fury over their complaints against him at Bukhara, he despaired. Askar Ali Khan argued for a southerly route through the mountainous Hazara country which would avoid Kunduz. Ghulam Hyder Khan and the rest preferred to chance their luck with Murad Beg again. Trebeck seems to have taken little part in this discussion. In November, as much from despair as disease, he too died.

Moorcroft had once referred to what a later generation called 'Kunduz fever' as a fatal disease 'not exceeded by the yellow fever of America or the fever of Walcheren'. Several of his servants including Mir Izzet Ullah had suffered from it during their detention by Murad Beg. It is not inconceivable that all three men, as they once again approached the fatal area, succumbed to the same disease. On the other hand, this was one of the most lawless areas of Asia.

In the light of what happened to European travellers in the area over the next twenty years, the death of three men in such a short time, all from natural causes, stretched credulity beyond belief. It was not just that it would have been an extraordinary coincidence, but that Moorcroft's party with their considerable resources were such an obvious prey. Few native caravans emerged from the area unscathed, let alone one led by Europeans. There was also the implacable hatred of Murad Beg, who may well have had wind of the proposal to by-pass Kunduz and was quite capable of wreaking his vengeance in the petty states beyond his frontier.

Moorcroft's travels had taken him a long way from the Western Himalayas. But the mysterious circumstances of his death lent to the already puzzling character of his travels a dash of romance. He became a legendary figure. Every traveller in the mountains and deserts between India and Russia added his stone to the cairn of 'the unfortunate Moorcroft'. The old vet was dead and yet his presence for decades haunted half Asia from Tibet to the Caspian. Wolff in 1831 met one of his employees who insisted that Trebeck and Guthrie, if not Moorcroft, had definitely been poisoned. Burnes, in the same year, was the last and only traveller to find his grave. Dr. Lord in Mazar-i-Sharif in 1837 recovered some of his books and accounts, and managed to thank the incredulous Pirzada at Taliq-an. ('Is it really a fact that this is known in Feringhistan?' asked the Pirzada. 'Wulla billa,' declared Lord, 'the very children repeat the name of Syed Mohammed Khan, the friend of the Feringhis.' 'God is great', quoth the Pirzada.) Gardiner met a man north of Kunduz with a compass and a map which had belonged to Moorcroft. French missionaries in Lhasa claimed, mistakenly, that Moorcroft had lived there for twelve years and died on his way back to Ladakh. In 1835, Henderson found that Moorcroft's garden in Leh was still being tended by the faithful Ladakhis in anticipation of his return. In Kashmir and Peshawar, Bannu and Rajaur testimonials written by Moorcroft were still being proudly shown twenty-five years after his death. His signature was said to be scratched on the wall of one of the famous caves at Bamian, and there was a shepherd just over the Bara Lacha who claimed that the flock he was grazing was Moorcroft's.

As the legend grew so did his reputation. The next fifty years were to prove this extraordinary man right in even his wilder speculations. It is almost uncanny the way events fell into place as he had antici-pated. Kangra, Kulu, Lahul and the whole of the Punjab including

Peshawar became British territory. Ladakh fell to the Sikhs, as he had warned, but later came under British protection as he had hoped. Bukhara fell to the Russians, while Eastern Turkestan and the mountains beyond Ladakh became the scene of frantic Anglo-Russian rivalry. Indo-British trade with Yarkand via Leh resulted in the formation of the Central Asian Trading Company, perhaps just the venture for which he had secretly planned, though it came too late to break the near monopoly gained by the Russians. He even advocated a permanent British representative in Leh and foresaw the possibility of a Muslim rising against the Chinese in Eastern Turkestan, both of which came to pass. The story of the exploration of the Western Himalayas is to a considerable extent the story of Moorcroft's dreams and fears becoming reality. Had all his recommendations been promptly adopted, the course of exploration and political penetration might have been run by 1830. But, by the government at least, they were, for the time being, ignored.

3. My Beloved Maharaja

Of the six major cities of the Indian subcontinent Lahore is now the least important. Beneath a grand Islamic skyline, all minarets, mosques and palaces, the narrow streets are as animated as ever. The salmon sandstone and white marble of the past are obscured though not shamed by the plate glass and cast concrete of today. Yet something is missing. There is industry and there is wealth but there is no longer dominion. Lahore has become a provincial backwater. It lies too close to Pakistan's vulnerable Indian frontier to have more than regional standing. It is still the capital of Punjab but Pakistan's Punjab is just a small province. Under Ranjit Singh it was an empire.

When Moorcroft was carried there in his *jampan* through the breathless summer nights of 1820 this empire was approaching its zenith. In 1818, with the capture of Multan and Peshawar, it reached its southern and western limits. The following year saw the conquest of Kashmir, later to be followed by the subjugation of Ladakh and Baltistan. The rule of Lahore then stretched from Tibet to Afghanistan and down the Indus to the deserts of Rajasthan and Sind.

Perhaps by Indian standards it was not such a vast empire. It could hardly compare with the conquests of Akbar or the sprawling dominions of Aurangzeb. Yet it was strong, more cohesive and stable than many a larger conglomeration of lands acknowledging one rule and, militarily, more formidable than anything the British had encountered in India. The Punjabis whatever their religion were, and still are, born soldiers; their physique and history bear witness to it. Sikhism, a comparatively new and suitably military religion, had become well established in the region. It gave to the non-Muslim population a crusading zeal comparable to that of the Mohammedan *jehad*. All that was needed was a leader who could unite the leading Sikh families, harness their Punjabi militarism and promote some sort of national identity in a land where such a thing was unknown. History rose magnificently to the occasion; there appeared Maharaja Ranjit Singh.

From the time of Moorcroft's visit till the Maharaja's death in 1839, no traveller from India could reach the Western Himalayas without first getting permission from Lahore. It was not the least of the obstacles the explorer faced, nor was the encounter with Ranjit Singh less revealing and colourful than the grimmer hazards of mountain travel. As a result of Moorcroft's indiscretions in Leh and his questionable activities in Srinagar, the Maharaja was highly suspicious of people professing an interest in the mountains. Kashmir was his prize possession. He guarded it jealously and had little confidence even in his own representatives there. The Honourable East India Company was equally suspicious of would-be travellers. Moorcroft had demonstrated how easily political complications could arise; the result was that the whole area was now out of bounds. Freelance travellers were strongly discouraged and officers of the Company were refused permission to cross the Sutlej as a matter of course.

For nearly ten years this situation effectively closed the Western Himalayas. No doubt the tardiness with which Moorcroft's *Travels* were brought to press was partly due to anxiety about the encouragement they might give to prospective travellers. Nevertheless, both in India and Europe, curiosity about what might lie within and beyond 'the snowy range' was growing. Three brothers from Aberdeen, James, Patrick and Alexander Gerard, made their speciality the small corner of the range which was directly accessible from British controlled territory. This lay east of Simla between Moorcroft's 1812 route over the Niti La and his 1820 route through Kulu. Any penetration into Tibet proved impossible; they were consistently turned back. But their descriptions of high altitude travel, the first of their kind, were quickly published. They reported peaks of close on 30,000 feet and themselves with the most primitive equipment reached heights of over 17,000 feet. It was under their guidance that Victor Jacquemont got his first taste of the Himalayas and was thus encouraged to present himself at Lahore in 1831 with a request for permission to visit Kashmir.

Victor Vinceslas Jacquemont was a French botanist. He had come to India to make a collection of flora and fauna for the Jardin des Plantes of Paris, a plausible story supposedly introducing a diligent and prosaic man of science. Had this been all he would have got little change out of Ranjit Singh. Nor would he have made much impression on Lord William Bentinck, the British Governor-General, whose good will he had first to solicit. But Jacquemont certainly wasn't prosaic. A dilettante and a wit, the darling of the

Paris salons and the friend of Prosper Merimée and Stendhal, he had dressed himself from head to toe in black and, armed with a bulging wallet of introductions, set foot, tall and twenty-eight, in Calcutta in 1829. Here was a very modern young man indeed. His ideas about the world were as liberal as those about himself were romantic. He is so detached from the Indian scene that he trips across it like a visitor from another planet. And so totally unexpected is he that one can understand how both Calcutta and Lahore were taken by storm.

After a turbulent but, to the would-be romantic, almost obligatory love affair with a celebrated opera singer he had sought solace and fulfilment in travel, first to America and now to India. The collection of natural history specimens was a convenient excuse. He pursued it with the sporadic enthusiasm of an eccentric's hobby; he was no mean naturalist but he found it hard to keep his eyes on the ground. Tibet, Kashmir and the Himalayas, these were already names to conjure with and, although their flora was hardly representative of India as a whole, he relished the possibility of being able to write to his Paris friends from such romantic places. Besides, the salary he received from the Jardin des Plantes was wholly inadequate. Of necessity his travels took him either to where the hospitality was warmest or else to places so remote that his frugality could pass for unavoidable hardship. He came to enjoy roughing it on milk and pillao one day and the next sitting down at the groaning board of a new British host.

This was the pattern of his journey as he sponged his way up country from Calcutta to Benares, Agra and Delhi. Probably it was the chance of more flattering attention, political gossip and square meals in Lord William's company that took him to Simla in 1830. At short notice Bentinck had to forgo his trip there that year, but Jacquemont's visit was not wasted. In Kennedy, the political agent to whom Moorcroft had first applied for arms for his private army, he found the perfect host, 'the first among artillery captains in the whole world'. Kennedy's job was to watch over a number of small hill states, which were south of the Sutlej and therefore subject to British protection. In practice he had almost unlimited authority and after an hour or two's work in the morning was free to devote himself to entertaining his guests. Simla's growth owed much to Kennedy's style. For Jacquemont the Perigord patés and truffles had to be specially ordered, but there was always a dinner lasting from seven-thirty till eleven. The visitor drank 'Rhine wine or claret or nothing but champagne, with Malmsey at desert, while under the pretext of the cold climate the others stick to port, madeira and sherry'. It was still a bachelor society in those days and after the strawberries and mocha coffee came the dancing girls. Jacquemont was reminded of Capua.

As well as the Gerard brothers here he met Dr. Murray from Ludhiana, the British listening post for Lahore. In his medical capacity Murray had been to the Sikh capital and was a good friend of General Allard, one of Ranjit Singh's French officers. News that the Sikhs employed ex-Napoleonic officers in positions of the highest authority suggested to Jacquemont a possible entrée to Lahore and thence perhaps to Kashmir and beyond. It was an opportunity not to be missed. Murray wrote to Allard, Allard invited Jacquemont and Jacquemont wrote off to Bentinck for a letter of introduction to Ranjit Singh.

This put the Governor-General in a quandary. The Frenchman was obviously a special case. He was not an employee of the Company nor was he British. To refuse him a recommendation would lead to bitterness and, possibly, embarrassing recriminations. But to give it might suggest to the Sikhs that he enjoyed some special relationship with the Company and that they supported his design to explore the mountains. Relations between the two strongest powers in India were extremely delicate. In most directions their interests were diametrically opposed. The so-called Anglo-Sikh alliance was the keystone of British India's relations with her neighbours yet, on both sides, it rested on little more than a fear of the consequences of a trial of strength disguised by extravagant and oft-repeated protestations of friendship. It was really more like a confrontation where the flicker of an eyelid on one side might send a hand flying to a gun on the other. In such an atmosphere a mishandling of Jacquemont's request could have dire consequences.

Bentinck was the soberest of Governors-General, a man dedicated to administrative and moral reform whose external ambitions were simply to maintain the status quo. He must have recalled the go-getting, empire-founding outbursts of Moorcroft with a shudder of horror. On the other hand he had a real affection for the dazzling young Frenchman, he rightly judged that he was not a political hoodlum, and he wanted to help him. He also observed that Jacquemont had a weakness for flattery. The personal letter of introduction to Ranjit Singh was, in the end, not forthcoming, but before Jacquemont could give vent to his disappointment he was introduced to the Sikh representative in Delhi in such flattering terms that he could scarcely keep a straight face. He was to continue as a freelance traveller but as 'the Socrates and Plato of the age' he could be sure of an impressive reception.

On March 2nd 1831 Jacquemont, well primed by the officers at Ludhiana, crossed the Sutlej. By contrast with Moorcroft he travelled light; just a cart, a couple of camels, three or four servants and the spirited little horse which had carried him from Calcutta. For the first few days he dispensed with the horse. For the most delicate of reasons he found the elephant provided by his Sikh escort more comfortable. 'The confidence of lofty souls meets with a poor reward at times. . . . I hope soon to have forgotten the dancing girls of Ludhiana.'

Often a preposterous figure, Jacquemont eventually wins one's respect. Few travellers have found their own misfortunes so amusing or been so candid about them. There had already been the celebrated episode of his stomach pump, which had made him the laughing stock of India. Like most travellers he had a vivid horror of gastric upsets and his own personal preventative. The daily clyster is by any standards a drastic precaution, but Jacquemont swore by it. So desperate was he when the vital intrument was stolen in Patiala that he confessed the loss to his blushing British friends. Descriptions and drawings were circulated to the Patiala authorities while the dispensaries of Upper India were scoured for a replacement. Eventually the original was found—no doubt the thief had mistaken it for a French hookah. The news was published in the Patiala court gazette and official notification of its recovery followed via the political department in Delhi. It was indeed the most celebrated syringe on diplomatic record.

Discomforts apart, the journey to Lahore went well. Jacquemont had been led to expect that as a guest all his expenses would be covered and that he would receive a small *nazzerana* each day. This was a cash offering which, like all such gifts, any servant of the Honourable Company had to pass on to his employers. Not so Jacquemont. And nor by his standards was the daily bag of coins so insignificant. A hundred and one rupees a day was two hundred and fifty francs. In the nine days it took to reach Lahore he would be richer by over two thousand francs, a third of his annual salary from the Jardin des Plantes. 'If Ranjit Singh feels obliged to treat his guests like this I can understand why he is not anxious to receive visitors . . . [he] has arguments that would reconcile me to the pace of a tortoise.'

Allard with a party of European officers met Jacquemont outside Lahore. They embraced with gallic fervour and in the general's four-horse barouche galloped off to the Shalimar gardens. '. . . we alighted at the gate of a delicious oasis. There was a great bed of stocks, irises and roses, with walks bordered by orange trees and jasmine beside pools in which a multitude of fountains were playing. In the middle of this beautiful garden was a little palace furnished with extreme luxury and elegance. This is my abode. Luncheon was awaiting us in my sitting room, served on solid silver.' In the evening there came presents of Kabul grapes and pomegranates and a purse containing five hundred rupees. Dinner was served by torchlight. It was 'exactly like a palace from *The Thousand and One Nights*'.

The following day Jacquemont had his first of many audiences with Ranjit Singh. Each had the highest expectations of the other; and neither was disappointed. Discarding what he imagined was a native disguise—green silk dressing gown, Indian trousers and a wide-brimmed black fur hat—Jacquemont appeared in his sober Paris suiting. He had no need of outlandish clothes. Tall and thin with cherubic countenance, vague grey eyes, a mass of chestnut curls and a new but imposing red beard, he looked every inch the romantic hero. The Maharaja liked to be surrounded by beautiful people. His most powerful courtiers, the brothers Dhian Singh and Gulab Singh, owed their elevation to their good looks; the newcomer was a decorous addition.

To live up to his billing as 'the Socrates and Plato of the age', Jacquemont tried to be as aloof and condescending as possible. But the Maharaja would not be put off. 'His conversation is a nightmare,' declared the Frenchman, 'He asked me a hundred thousand questions about India, the English, Europe, Bonaparte, this world in general and the other one, hell and paradise, the soul, God, the devil and a thousand things beside.' Jacquemont could never resist the chance to air his views and his famous charm soon thawed the frigid pose. Its combination of effortless flattery, eloquent wit and an apparently irresistible tone of voice convinced the Maharaja that here indeed was a remarkable man, a veritable demi-god. Whatever his advisers might say this was no Englishman in disguise. Nor would the

English dream of sending such a fluent and interesting young man as a spy. He could bring nothing but credit to Lahore and might go wherever he wished, even to Kashmir.

As for Jacquemont it was a case of love at first sight. 'My beloved Maharaja,' he called him. First sight of Ranjit Singh was not usually prepossessing. 'Exactly like an old mouse with grey whiskers and one eye,' was how Emily Eden described him seven years later. He was very small, almost a dwarf, and even his good eye was failing. His face was pockmarked, his speech slurred and his bearing ungainly. Yet there was no denying an impression of extraordinary energy. His solitary eye blazed with animation, his hands were alive with expression and there was no finer horseman in the whole Sikh army.

Here was a man who had moulded history, like Napoleon a self-made emperor. From obscure origins he had united the Sikh clans, repulsed the Afghans and started to build. Aged eighteen he had captured Lahore and been recognised as chief of the Sikh leaders. At twenty-one he proclaimed himself Maharaja of the Punjab and five years later was negotiating with the British on an equal footing. His courage was unquestioned; it attracted a touching loyalty from his old comrades-in-arms and the empire depended, perhaps too much, on his personal magnetism. Undoubted too were his political insight and opportunism; the British tended to call them avarice and cunning. He could neither read nor write and sometimes for days on end he would conduct no public business. Yet little escaped his attention and few were the men he misjudged.

To be flattered by such a man was praise indeed. Jacquemont revelled in it. Above all he valued the Sikh's confidences. Ranjit Singh's two greatest passions were horses and women. Moorcroft had found common ground on the former. Jacquemont was clearly better informed about the latter. A life of unremitting sexual indulgence had left the Maharaja impotent at fifty. He described his problem as a weakness of the digestion and in case his meaning was not quite clear he proceeded to demonstrate. '. . . the old roué sent for five young girls from his seraglio, ordered them to sit down in front of me and smilingly asked me what I thought of them. I said that in all sincerity I considered them very pretty, which was not a tenth of what I really thought.' The girls were then put through their paces. Even Indian singing had its appeal in such absorbing circumstances. But the Maharaja remarked sadly that he had a whole regiment of these exquisite Kashmiri girls. He would review them mounted bareback on chargers, a combination calculated to excite his deepest emotions. He would ply them with liquor till their performances became naked revels and their frolics savage cat fights. But all to no avail. His 'digestion' failed to respond. In fact the problem had become a matter for general concern. Unresponsive to the women of his own harem, he had taken to trying his jaded prowess on those of other men or even public prostitutes. The streets of Lahore witnessed scenes unprecedented as he drove through them in an enormous state carriage fitted with a verandah which accommodated twenty *nach* girls. Live performances were also given on elephant back, but it was going too far when the girl concerned was a Moslem courtesan and the particular diversion what Jacquemont calls the 'honteux pis-aller des vieux libertins'.

The Frenchman prescribed 'cantharides', evidently a species of beetle, dried. They were reputed to have aphrodisiac qualities and Elphinstone had won much acclaim with them on his mission to Kabul. Whether Ranjit Singh tried them is doubtful. He was usually too suspicious to do more than dose his courtiers and await their reaction.

Jacquemont spent two weeks in Lahore. Like Moorcroft he wanted to get away before the impossible heat of May. It was a measure of his success that instead of prevarication he received generous assistance, five hundred rupees on parting, five hundred more en route and two thousand awaiting him in Kashmir. By then his grand total would be more than two years' salary. There were other presents, Kashmir shawls, silks and jewels, camels to carry his baggage and a secretary to see that the Maharaja had a full report of his progress. Allard checked through his equipment, supplying any additions he thought necessary, and threw a grand and 'plus galante' farewell party in the Shalimar gardens. The Kashmiri girls were more lovely than ever. One in particular attracted his attention. She was beautiful by any standards, a real 'princesse d'Opera'. In the half-dark, while the servants were lighting the saloon and the rest of the party were dispersed through the garden, who should come to him but the lady herself. Allard's generosity was on a par with Kennedy's. And the ladies of Lahore did not 'betray the confidence of a lofty soul'. He left, on horseback, on March 25th.

His immediate destination was Kashmir but where Jacquemont intended to go from there is not clear. From Simla in 1830 he had attempted, like the Gerards, to cross into Tibet. In his own book he had succeeded. It is hard to follow his exact route but he claims to have crossed the frontier undetected and penetrated a day's march before being stopped by 'Chinese' guards. Thinking their reluctance to dismount was lacking in respect, he summoned one for a parley and promptly grabbed his pigtail and unhorsed him. It made an excellent story for home consumption and he had withdrawn to British territory well contented. At one time he certainly planned to continue his Tibetan campaign from Kashmir but it seems the British must have talked him out of this. The plan was now to be the more modest one of proceeding from Kashmir to Ladakh, and thence back to Simla either via Spiti and Bushair or by Moorcroft's Kulu route. In the event neither was possible. He was lucky to get as far as Kashmir safely.

Trouble started as soon as he reached the foothills at Mirpur. Ranjit Singh had promised him mules and porters to take over from the camels; neither materialised. The villagers fled before him and even his Sikh escort started to melt away. Supplies were hard to come by, the water was so muddy it looked like chocolate and when the rain was not dissolving the labels on his geological specimens the temperature soared to 93°. A man like Moorcroft would have patiently bided his time but Jacquemont was not used to fending for himself. Till now his path had been smoothed by the reassuring authority of a Bentinck, a Kennedy or a Ranjit Singh. Here it seemed that though still within the territories of Lahore it was every man for himself. Cursing, coercing and, as a last resort, considering paying his porters, he pushed on ahead with a scratch complement.

April 22nd dawned full of promise. No porters had absconded during the night, the weather was fine and cooler, and they were under way in good time. Jacquemont's horse was lame but the path was anyway too steep and rough for safe riding. He was in fact ascending an almost perpendicular mountain. The top was flat and on it stood a hill fort. After an hour's stiff haul he was over the lip and looking on a fine sward of grass with the fort in the middle. Groups of ragged soldiers armed with swords and matchlocks were dotted about and Jacquemont's party lay resting in the shade of a giant fig tree. It was very picturesque but too early for breakfast. The order to proceed was given. Nothing happened. The men weren't resting, they had been ordered to stay there. The governor of the fort turned out to be a half-starved bandit called Nihal Singh, and Jacquemont was already his prisoner.

This Nihal Singh deserves a better press than he gets from Jacquemont's biographers. He had been appointed governor of the fort by the Maharaja and right faithfully had he held it, even against the demands of Gulab Singh, the Maharaja's most powerful deputy in the mountains. Like most such governors he would have been told not to surrender his charge to anyone except Ranjit Singh himself. But Gulab Singh had seen to it that the obstinate hill man suffered for his intransigence. For three years he had received no pay. His men were starving and a prestigious hostage was just what he needed. Jacquemont had already met and liked Gulab Singh, but history has revealed him as a cruel and scheming autocrat. At the time he was busy carving out for himself a semi-independent kingdom in the mountains, the power base from which, after Ranjit Singh's death, he was to manœuvre for Kashmir. In the context of his plans this fort, commanding one of the main routes from Lahore to Srinagar, was an essential acquisition.

Cheered on by the armed rabble who now jostled and threatened, Nihal Singh explained the situation to his unsympathetic prisoner. He did not yet realise that the cruellest blow of all was to have picked on a man like Jacquemont as hostage. For the Frenchman it was to be his finest hour. The desperadoes were shaking the ash from their already lighted matchlocks. His own men were miserable with fear and closely guarded. Help, in the shape of Gulab Singh, lay many days' march away and the fort was in any case impregnable. His only weapon was bluff. He affected an arrogant, world weary and devastating unconcern; to the romantic young man it was second nature. Taking Nihal Singh firmly by the arm he sat him in the shade of the fig tree. Then he had a chair brought for himself. Undismayed Nihal Singh started again to explain; the traveller did not seem to understand that he was a prisoner. Jacquemont interrupted. He summoned his servant and ordered a drink. It was a long time coming and then he felt too hot to conduct business. His parasol was laboriously erected and another servant cooled his master with a peacock fan. At last Jacquemont turned his short sighted gaze on his captor. In the most confidential tone he pointed out the enormity of the crime contemplated and the hopelessness of Nihal Singh's position should he persist with it. Nihal Singh reconsidered. Perhaps relieved that he was not yet fully committed and a little unsure of his peculiar prey he suggested a generous compromise. He would just hang on to the party's baggage. Jacquemont exploded.

'What? Travel without my tents? Without my furniture? Without my books? Without all my clothes? I who change twice a day!'

Then carefully studying his watch he ordered breakfast. His servant explained that nothing was ready. Even milk was not to hand.

'Don't you hear?' bellowed Jacquemont at Nihal Singh, 'The lord wants some milk. Send to the neighbouring villages as quickly as possible so that it may be brought at once.'

It was his masterstroke. Nihal Singh was confused. What sort of a man was this who used not the royal 'we' but was so grand he talked of himself in the third person? He hesitated, then gave the order. A small party set off for the milk. Jacquemont waited till they were almost out of earshot before ordering them back. It must be cow's milk, not goat's or buffalo's, and they were to watch the cows being milked with their own eyes. He was just rubbing it in. The victory was his. ''They were crushed by my disdain'; it had also taken a lot of courage.

In the end Nihal Singh got just five hundred rupees and even this he would only accept on the condition it was given as a present. For now he was Jacquemont's most grateful friend and most faithful servant. He offered an escort, a doubtful security, and whispered a request for a bottle of wine. Overcome with relief, Jacquemont still had the presence of mind to give him a bottle of the Delhi arrack, which he used as a preservative, rather than any of his vintage port. Later he sent Ranjit Singh a full report of the affair and received ample compensation. Nihal Singh was captured and imprisoned at Jacquemont's pleasure. Though the Maharaja was celebrated for his leniency, this was an exceptionally humane sentence. It may have been in deference to European sensibilities. More probably it can be taken as further evidence that poor Nihal Singh had a better case than most 'brigand chiefs'.

Jacquemont pressed on for Kashmir. In Punch he collapsed spitting blood from what he took to be a chill. Sixty-five leeches fished from nearby rivers and applied to his chest and stomach seemed to cure the trouble. He crossed the Pir Panjal by the easy pass above Uri and reached Srinagar on May 8th 1831. He claimed to be the first European since Bernier to visit Kashmir undisguised. True, Forster had tried to hide his identity and de Koros, for convenience rather than necessity, had also worn native dress. But Moorcroft and Trebeck certainly stuck to their European coats.

It may be significant that Jacquemont could be wrong about this and yet surprisingly well informed about later episodes in Moorcroft's travels. In 1831 these were still unpublished. But at Kennedy's house Jacquemont had met and become close friends with William Fraser. So much so that Fraser, a renowned misanthropist, applied for permission to accompany him to Kashmir. This was the man amongst whose papers Moorcroft's last journals were found. It looks as if in 1831 these were already in his possession. Jacquemont is quite clear about Moorcroft's dying in Central Asia and the causes being 'a putrid fever, or else a dose of poison or even gunshot'. But of his doings in Kashmir, chronicled in journals which Fraser never possessed, he is poorly informed. A case in point may be discerned in his handling of approaches from Ahmed Shah, ruler of Baltistan. He seems to have imagined it was Moorcroft's overtures to this Raja, rather than to the Ladakhis, that earned him Ranjit Singh's displeasure and the Company's censure. (Moorcroft was in touch with Baltistan but it had not led to any repercussions.) Jacquemont's mistake, though it was to cost Amhed Shah dear, had the effect of bringing the existence and predicament of this remote mountain kingdom to the attention of the outside world.

Baltistan lies due north of Kashmir along a stretch of the Upper Indus and between the Great Himalaya and the Karakorams. The name derives from the people, the Baltis, who doubtless were the Byltae of the classical geographers. It is also called Skardu (Skardo or Iskardo), after the principal fort, and sometimes Little Tibet as distinct from Middle or Western Tibet (Ladakh) and Greater Tibet (the country which still bears the name). Jacquemont always calls it Little Tibet; it was the more romantic of the three. No European had ever been there and though he hoped to explore towards it, he showed no inclination to be the first. Already he had decided that Ladakh was beyond his means and was quite content making short botanical forays round the Kashmir valley.

Returning from one of these he found, instead of the expected courier with a batch of letters from France, a man claiming to be an emissary from Ahmed Shah. It was flattering to find that his fame already extended to the remotest valleys of the Western Himalayas and Jacquemont glowed at the oriental compliments of his unknown admirer. He was less happy when it transpired that Ahmed Shah had taken him for a British agent. This rather diminished the value of the compliments and threatened to prejudice his good relations with Ranjit Singh. The agent affirmed his master's attachment to the British, his dislike of the Sikhs and his expectations of what 'Jackman sahib' could achieve with a couple of British regiments. Recalling what he thought he had heard of Moorcroft's indiscretions Jacquemont made sure that Ranjit Singh had a report of the affair. The Sikhs were after all not just his hosts but his paymasters. Much as he resented the accusation, Jacquemont in Kashmir was in effect the Maharaja's spy. He returned Ahmed Shah's compliments and insisted that his interests were purely scientific.

Two months later, while botanising in the north-western corner of the Kashmir valley, he was again visited by agents from Baltistan. 'I cannot make out what these people want of the English,' he wrote. Ahmed Shah had the reputation of being a model ruler, beloved by his subjects and feared by his neighbours. His kingdom was safe enough from the Sikhs; it was too poor to attract conquerors and quite inaccessible. '. . . for all my diplomatic genius I cannot make it out'. Since there was no way the Sikhs could injure Ahmed Shah, he again sent a full report of the meeting to Lahore. But he was wrong. These reports were to have serious repercussions. The Sikhs could reach Baltistan and, armed with this evidence, were inclined to do so. The first of several attempts to invade the country was made in 1833, and for the next six years the fate of Baltistan was to be a decisive factor in the exploration of the mountains.

Jacquemont spent four months in Kashmir. He amassed a considerable collection of plants, geological specimens, stuffed birds and preserved fishes; there was even a pair of live antelopes sent by Ahmed Shah. Yet he was unhappy. It was usual for the romantic hero to carp and boredom was second nature to him, but this was something more. Earlier he had written, 'sometimes I cannot believe that it was I who did this or that or went here or there . . . at times I doubt my own identity'. No one who has been alone in strange surroundings will question such sincerity. The symptoms speak for themselves. He was lonely and very homesick. Had there been a post office he would have been hanging listlessly about outside it. He himself wrote reams, in 1831 346 substantial letters as well as his weighty journal. His regular correspondents were friends and family in France whose letters came round by the Cape to Chandernagore, the French colony near Calcutta, then up country to Wade, the British agent at Ludhiana, then to Allard at Lahore and finally across the mountains by special courier. In July he received the first news from home in twelve months. 'Now I am happy . . . my hand shakes, my ideas are in confusion.'

This euphoria was short lived. In July and August he found the heat stifling and was one of very few who have actually longed for the scorching winds of the plains. He was thinner than ever and seriously worried about his health. *Lalla Rookh* bored him, the Kashmiris depressed him and the scenery was disappointing. As for the women, they were hideous; the handsome ones were all exported to the Punjab. It was not till he received what he chose to regard as a summons from Ranjit Singh to return that his favourite aphorism reappears—'All is for the best in this best possible of worlds'.

With an escort of sixty cavalry—he was not to be at the mercy of another Nihal Singh—and some fifty porters loaded with his specimens he left Kashmir on September 19th. The route this time was via the Pir Panjal pass, the one used by Moorcroft, and thence to Jammu and Amritsar. Here he found the Maharaja who was on his way to a celebrated meeting with the Governor-General at Rupar. Jacquemont had resolved not to accompany him—'a mere Plato would be lost in the dust at the meeting of two such immortals'. Already the Sikh court was in a high state of excitement, yet Ranjit Singh found time for a protracted farewell. He tried to persuade Jacquemont to stay, offering him no less than the viceregency of Kashmir, an inducement which the Frenchman correctly judged to be purely figurative. At Hoshiarpur they finally parted.

> Our last interview was long and very friendly; Ranjit lavished a thousand caresses on me; he took my hand and shook it several times at my well aimed broadsides of flattery into which I infused, spontaneously, a deal of feeling.

Moist-eyed but well pleased, both with himself and with Ranjit's parting presents, Jacquemont left the royal camp. He had understood the Maharaja better than almost any other European. They were an ill-assorted pair, the young, elegant, educated Parisian and the old, ugly and illiterate emperor. Yet as seen through the other's eyes both acquire stature. To a man from whom he correctly surmised that there was little to fear, the Maharaja displayed a genuine warmth and loyalty. Jacquemont reciprocated. He rejoiced in Ranjit's patronage but, to his credit, never abused it.

*　　*　　*

On the occasion of his thirtieth birthday in Kashmir Jacquemont had written, 'I think a man must be rather foolish to allow himself to die at thirty.' A year later in Poona, when ill health was getting the better of him, he claimed that 'a traveller in my line has several ways of making what the Italians term a *fiasco*, but the most complete fiasco of all is to die on the road'. A few months later he did just that. He was thirty-one.

After recouping with Kennedy at Simla and with Fraser at Delhi he had set off during the hottest months of the year for Bombay, a nightmare of a journey through the wastes of Central India. It demanded a different sort of courage to the sangfroid he had deployed against Nihal Singh, something more like grim determination. He had it in good measure; but the journey was too much. In the jungle just north of Bombay he was taken ill with fever which produced an abcess of the liver. For some weeks he suffered in appalling pain before dying peacefully in Bombay on December 7th 1832.

Much about Jacquemont is infuriating, his vanity, his posing and his selfishness. He is too glib, too determinedly witty, to be very funny. And yet one must admire him. He was no fool and no coward. He was the most unlikely of explorers, yet he was the first and by no means the least in the long tradition of Himalayan naturalists. In his death a real dignity emerges. Mérimée called him 'a stoic in the real sense of the word', a man who 'took as much pains to conceal his emotions as others do to hide their evil propensities'. In his last letters the mask starts to dissolve. It does not reveal the shallow, ordinary character one suspects under all those airs and affectations. There emerges a truly courageous and generous soul. He studies the progress of his death with a calm but intense interest. He reassures his family lest they think him dying lonely and friendless. In fact he is glad they are not present; he warmly commends the Englishmen who are looking after him and fears that the grief of his father and brothers would be too much for him to bear. To the very last 'all is for the best in this best . . .' but it is impossible to read on without tears.

4. A Heathen King

In the spring of 1832, while Jacquemont was making that gruelling journey from Delhi to Bombay, another European was toiling east through Afghanistan. He had passed through Bukhara, the first visitor there since Moorcroft, and was now retracing the steps of his unfortunate predecessor towards Punjab and the court of Ranjit Singh. He, too, had his heart set on Kashmir and the lands beyond the mountains. And though as odd a character as ever set foot in the Himalayas, his chances of getting there were not to be underrated.

But first he had safely to negotiate Afghanistan. At Khulm he escaped the attentions of Mured Beg by hiding in his room at the caravanserai and slipping out before dawn. It was altogether easier for him than it had been for Moorcroft. He had just three servants and negligible baggage. The clothes he wore were not a conscious disguise. They were just the sort of travel-stained cottons that anyone who had been wandering about the Middle East for five years might have been reduced to. Provided he kept his mouth shut, he was as safe as anyone in this lawless corner of Asia.

Four days south from Khulm, at a place called Doab, he left the lands of Murad Beg and entered the mountainous country of the Hazaras. These are a Mongol people, descendants of Genghis Khan, and every bit as wild as their Uzbek and Afghan neighbours. Something about the traveller, perhaps the tattered book which he always carried under his arm, drew their attention. They demanded to know his name.

His problem was not one of language. He could speak any number, including Persian, the *lingua franca* of the region. The difficulty was that to him it was a matter of principle not to tell a lie. He could mislead but under no circumstances would he deny his identity or his purpose. A week previously he had told the governor of Mazar-i-Sharif, a man who was sworn to kill every European and whom he believed to be personally responsible for the deaths of Moorcroft and his companions, that he came from Malta. He was not Maltese but back in 1830 it was from Malta that he had begun this particular journey. The governor had never heard of Malta; he wanted to know where it was. By way of an answer he was told that there, instead of boats made of skins, they had steamships. These made a distinctive noise which the traveller imitated and in one of them you could reach Malta from Constantinople in four days. As for the Governor of Malta he was called Ponsonby Khan, son of Bessborough Khan and his wife was the daughter of Bathurst Khan. This was too much for the Mazar chief. Appalled at his ignorance of this supposed Islamic sea-power, he had beaten a speedy retreat.

At Doab the stranger was hoping to use the same tactics. They wanted to know his name. He would tell them.

'Haji Youssuf,' he replied.

Youssuf is the Arabic Joseph and a *haji* is someone who has made the pilgrimage to Mecca. The Hazaras asked for the blessing of such a holy man. But they must still have been suspicious for they now turned to one of his servants and asked if the *haji* were really a Mohammedan. The servant said yes. Haji Youssuf said no.

'Why do you take the name of *haji* if you are not a Mohammedan?'

'Even the Mohammedans in Bukhara recognise as *haji* all Jews and Christians who have been to Jerusalem.'

'This is not the custom here among us. We are *kharijee* [i.e. schismatics; they were actually Shiah]. With us many things are not allowed which are allowed by other Mohammedans.'

'I could not know your usage for I have but just arrived among you; so all you can do is not to call me *haji*; I shall tell my people not to call me *haji*.'

'The mischief is done and therefore you must either say "There is God and nothing but God and Mohammed is his prophet" or we will sew you up in a dead donkey, burn you alive and make sausages of you.'

'There is God and nothing but God and Jesus the Son of God.'

It was as good as signing his own death warrant. The Hazaras assembled with their Mullahs in a nearby cave and prepared to pass formal sentence. The traveller asked for his writing box and in great terror, for he was not a brave man, wrote the following letter.

To Lord and Lady William Bentinck.

Dear Lord and Lady William Bentinck

The moment you read this you must beware that I am no longer in the land of the living; that I have been put to death. Give to my servants some hundred rupees for their journey and write the whole account to my wife, Lady Georgiana.

Your affectionate,

Joseph Wolff.

He confided the letter to his servants who were to travel with all haste to Ludhiana and give it to the first redcoat they saw. Then he bade them farewell and turned with the tattered Bible still under his arm towards the cave.

Joseph Wolff, that was indeed his name, was by 1832 used to persecution. Since the age of seven he had been angering people with his religious mania. And he was now thirty-eight. Born the son of a Bohemian rabbi, he had started asking awkward questions as soon as he could talk. At seven he was all for abandoning the Jewish faith in favour of Christianity. At eleven he wanted to be a Jesuit; an old aunt was so provoked she threw the poker at him. At seventeen he was baptised a Catholic, and at twenty-one he arrived in Rome to study for the priesthood. By then there was no town in Germany with the most modest academic pretensions from which he had not been expelled. Rightly he saw himself as the *enfant terrible* of the theological world, and it was just a matter of time before Rome too would reject him. In the event he was treated with great patience and retained a lifelong respect for Catholicism. But when, after eighteen turbulent months, he was still objecting to papal infallibility and flirting with protestantism, the Inquisition stepped in. He was 'rolled out of Rome' under guard and deposited in Vienna. Again he did the rounds of the German monasteries, trying even a flagellant order before he finally took up an offer of patronage from Henry Drummond, a wealthy English eccentric. He arrived in London in 1819.

It is no good looking for the modern equivalent of a man like Wolff. He was an extraordinary phenomenon and one not likely to be repeated. The twentieth century is too indifferent to religion and too sensitive about mental derangement. Such a man if not actually shut away would today be smothered by sympathy, his opinions curdling inside him. A shabby figure known to all at the public library, his greatest act of defiance would be a stream of never published letters to the newspapers. In the early nineteenth century society was less kind. Wolff was taken at his own value and made to pay for it. Instead of pity he met with ridicule. Instead of being ignored by the papers they pilloried him. And instead of being protected he was packed off to the wildest regions of Asia.

Yet he thrived on it. Wolff's life was a supremely happy one. Hardship, loneliness, exhaustion and ostracism were normal to him. Yet no man was ever further from despair. He was doing his life's work and at the age of sixty-five he could still tell the story of it with unadulterated satisfaction. Age modified some of his more outrageous ideas and caused him to regret his youthful intolerance. But he was still a little mad. He was also cantankerous, excitable and vain, a difficult man to get on with and a colossal embarrassment in public. It was, and still is, easy to be funny about him; less easy, a hundred and fifty years later, to understand how people could love him. But even Bentinck, whom Jacquemont described as an upright, simple and sincere administrator 'like a Quaker of Pennsylvania', doted on Wolff. And he was far from being alone.

The English have always had a weakness for eccentrics and Wolff liked what he saw of England and, even more, what he saw of the Church of England. Here was a Christian institution elastic enough to accommodate his unorthodoxy and rich enough to be interested in proselytising. As a converted Jew, Wolff's lifelong ambition had been to preach Christianity to the Jewish people. It was with this in mind that he had studied Hebrew, steeped himself in scripture and joined the Propaganda in Rome. Now at last he found a man, Henry Drummond, and an institution, The Society for Promoting Christianity among the Jews, both willing to support him. The latter sent him to Cambridge to continue his oriental studies and the former, two years later, packed him off to the Mediterranean on his first mission.

He was away for five eventful years. He visited almost every city of consequence between Teheran, Cairo and Constantinople, and claims to have been the first Jew since Christ to preach Christianity in Jerusalem. His diaries read rather like the daily call sheets of a conscientious salesman. 'Proclaimed Christ to so-and-so. Requests Bible in Kurdish.' He met with some rough treatment and converts were not easily won. But unlike previous Jewish missionaries from England, 'one of whom became a Mohammedan, another a thief and a third a pickpocket,' Wolff stuck to his guns.

On the day he arrived back in England he met at a dinner party Lady Georgiana Walpole, daughter of the Earl of Orford. That any woman should want to marry Wolff seems strange. In his own estimation he was a great favourite of the ladies but, even if true, this was not quite the same thing as being eligible. Besides Lady Georgiana was an altogether improbable conquest. She was passably handsome, of independent means and impeccably connected. To Wolff she became 'his darling angel in earthly shape', and six months later he married her. Neither seems ever to have regretted it. Shortly after he became a naturalised British subject. He could now count on a warm reception at any distant outpost of the Empire to which his travels might take him.

In 1827 the happy couple left England to resume Wolff's work. For three years he continued to tour the Middle East with Lady Georgiana never far behind. In Jerusalem, the scene of his earlier triumphs, he received a cold welcome and was poisoned in a coffee house. Lady Georgiana was there to nurse him, but it was galling to be an invalid and outcaste when he had expected to be the conquering hero. He was beginning to lose interest in the Jews of Jerusalem and to look further afield.

He returned to Malta in 1830. Lady Georgiana joined him there, now with their second child,* and they discussed plans to visit Timbuktu. It is not clear whether Wolff imagined he would find Jews in the Sahara but when a friend suggested Bukhara and Afghanistan, where there were not only Jews but remnants of the lost tribes, he readily agreed. 'Wolff shouted', he always writes of himself as plain 'Wolff' and he often shouts, 'Wolff shouted, "To Bukhara I shall go!"'' The Society for Promoting Christianity among the Jews wanted him back in London. Wolff wrote that he was on his way— via Bukhara, Tibet and Calcutta—and on December 29th 1830 said a sad goodbye to Lady Georgiana and set off for Turkestan and the Himalayas.†

Such a journey, if successful, promised to rival that of Marco Polo. The dangers were appalling but, as he now turned to confront the Mullahs of Doab, he could reassure himself with the recollection of many an equally hopeless situation. He had already been captured

* Their first child, a daughter, lived only a few months. The second, christened Henry Drummond Wolff, was to have a distinguished diplomatic career.

† This was the 1860 account of the inspiration for the journey. In the 1835 account he says that the idea came to him while still in Jerusalem. 'I said to my wife, "Bukhara and Balkh are much in my mind for I think I shall find there the Ten Tribes." "Well," she said, "I have no objection to your going there."'

by slave dealers and dragged across the desert tied to a horse's tail. He had always escaped and he was still going strong. As a traveller his experience was unique; there was yet momentum to carry him across the Himalayas.

The mullahs of Doab were seated in the cave with the Koran open before them. Wolff himself takes up the story.

> Wolff said, 'What humbug is that? You cannot dare to put me to death. You will be putting to death a guest.'
>
> They replied, 'The Koran decides so.'
>
> Wolff said, 'It is a lie. The Koran says on the contrary that a guest should be respected even if he is an infidel; and see here the great *firman* I have from the Khalif of the whole Mohammedan religion from Stamboul. You have no power to put me to death. You must send me to Mohammed Murad Beg at Kunduz. Have you not seen how little afraid of you I am? I have told the Afghans [with whom he had been travelling] that they should disperse and probably some of them have already gone to Kunduz.' When they heard the name of Murad Beg they actually began to tremble and asked Wolff 'Do you know him?' As Wolff could not say that he knew him he replied, 'This you must find out.' They said 'Then you must purchase your blood with all that you have.' Wolff answered 'This will I do. For I am a dervish and do not mind either money, clothing or anything.'
>
> And thus Wolff had to surrender everything. Oh, if his friends in England could have seen him then they would have stared at him. Naked like Adam and Eve and without even an apron of leaves to dress himself in he continued his journey.

For the next six hundred miles Wolff claims that he travelled without clothes. If his friends in England could have seen him they might have looked on 'an obese and dauntless Sebastian'. That was how *Blackwoods Magazine* described him facing his critics, and it certainly corresponds with the portrait we have of him at sixty-five. On the other hand they might have seen a man 'in stature short and thin' with 'a weak frame'. This was an earlier eye-witness and one can only presume that somewhere along the line he put on weight prodigiously. He was never indifferent to food, succumbing to pizzas in Rome and plum pudding in Cairo. A profile drawing of him in 1840 already reveals a double chin. Perhaps, too, one did not normally notice his figure. The 'strange and most curious looking' visage was enough to hold the attention. It was 'very flat and deeply marked with smallpox', the complexion 'that of dough' and the 'hair flaxen'. People were reminded of Luther except that the eyes were too disquieting. They 'roll and start and fix themselves most fearfully; they have a cast in them which renders their expression still wilder'.

Such was the apparition that now flitted across the mountains of Afghanistan. The Hindu Kush here is not the majestic range it becomes in the Western Himalayas but it is still an extensive area of desolate terrain with passes of eleven and twelve thousand feet. It was only April and on the high ground the snow lay deep. Tumbling into snowdrifts and running from the icy wind, Wolff kept on for Kabul. On the first of May he emerged from the mountains and came within a few miles of the city. He was now in a state of collapse; he was also penniless and still naked. It was unthinkable to enter the city in such a condition, and anyway his servants were not going to let him. He owed them money and they wanted to be sure that his credit was good before going any further. With no great expectations Wolff wrote for help to Dost Mohammed, the ruler of Kabul, and prayed to his Saviour.

Whenever a situation like this had arisen in the past, assistance had come in the shape of 'a British soldier sent to him by God'. In 1823, when he had been sick in Jerusalem, a British colonel had appeared from nowhere and nursed him over the worst and set him on his feet again. In 1824 the same thing happened in Baghdad when he stumbled into the city after receiving a 200-lash bastinado. In 1825 it was the Caucasus. Wolff, prostrate with typhus, lay dying beside the road when a guardian angel, again in colonel's uniform, swept him into his carriage and off to the sanitorium of a nearby monastery. 1826 was a blank but in 1827 it was no less a man than General Sir Charles Napier who rescued him in Cephalonia after being washed ashore, destitute, from a shipwreck. 1828, dysentery in Cairo, and a Colonel Felix relieved Lady Georgiana of the job of nursing him. And finally 1829, when having been attacked by pirates near Salonica he had drifted ashore and wandered about barefoot for two days, it was a naval lieutenant who came to the rescue with an advance of clothing and money.

This was just the help he needed now though the chances of a British officer being anywhere near Kabul were dim. None had been sent there since Elphinstone in 1808. But Wolff had great faith and, sure enough, within two hours there appeared men with a horse and clothing for the traveller and a letter. It was addressed to him in English.

The writer turned out to be Lieutenant Burnes of the Bombay army. He had arrived in Kabul the day before along with Dr. James Gerard, one of the three Aberdonian brothers whom Jacquemont had met in Simla. At the very least it was an extraordinary coincidence. Burnes and Gerard were en route to Bukhara, a journey which was to be the making of the former and the death of the latter. Gerard was now well into his fifties and already a sick man. But he welcomed Wolff and enjoyed the diversion when their house was turned into a debating chamber for all the Jews and Armenians of the city. Burnes was less enthusiastic. Probably he was a little piqued at having been beaten to Bukhara. A year later he was to do his best to expose Wolff as both a lunatic and a liar.

Wolff did not stay long. He told Mohan Lal, Burnes's secretary, that in Bukhara he had had a vision in which Christ told him that the valley of Kashmir would be the new Jerusalem. There were also persistent rumours that Kashmir was the site of the Garden of Eden and that the population belonged to one of the lost tribes. Jacquemont particularly had remarked on the physical resemblance between the Kashmiris and the Jews. Already Wolff was convinced that the Jews of Turkestan were of the lost tribes (also, surprisingly, 'the descendents of Genghis Khan, the Nogay Tartars and those called of the tribe of Naphthali'). He was less sure about the Kafirs of the Hindu Kush, but the Kashmiris sounded very promising. He could hardly wait to get there.

He passed through Jalalabad and Peshawar and crossed the Indus at Attock. After the savagery of the Turkomans, the bigotry of the Hazaras and then the wild Pathans of the Khyber, Punjab was a welcome change. He was greeted by kind respectful people dressed in flowing white cottons. Their hands were folded as they awaited his orders. It was the sort of reception he envisaged receiving at the gates of Heaven. If only Ranjit Singh were not just 'a heathen king' but the King of Kings; 'Well done,' he would say, 'thou good and faithful servant, enter thou into the joy of the Lord.'

Ranjit Singh was not much interested in religion but he was doing his best for Wolff. Lord William Bentinck, who had been warned of the missionary's approach by the British representative in Persia, had asked the old Sikh to look after him. Instead of Jacquemont's

1 In the Pir Panjal near the Dulchi Pass

2 William Moorcroft: supposed portrait

3 Joseph Wolff in.1840

4 Victor Jacquemont

5 Baron Carl von Hugel

hundred rupees a day he was getting two hundred and fifty and was free to make his own way towards Lahore. It was a striking contrast to his treatment in Afghanistan. He was entertained by Hari Singh, one of the greatest Sikh commanders, Kharak Singh, the heir apparent, and then Josiah Harlan,* an American who held the governorship of Gujerat, and Paolo Avitabile,† the Italian governor at Wazirabad. 'The kind Italian' was really a cruel libertine and the 'very interesting American' a two-faced scoundrel who played the Sikhs as false as he had the British before them. Wolff was never much of a judge of character and he was overwhelmed by these receptions. So much so that after leaving Avitabile he made what later turned out to be a fatal blunder.

Meantime, dressed in a European suit made for him in Peshawar and crisp new linen provided by the Sikhs, Wolff was carried in a palanquin to Lahore and then on to Amritsar where the Maharaja was spending the summer. On June 21st he was presented at court. An elephant was sent to fetch him and after passing through three courtyards he was ushered towards a figure which, with his feeble sight, he took to be a small boy. He was about to ask him if he was one of the Maharaja's *bachas* when he made out the long grey beard.

---

* Josiah Harlan was born in Pennsylvania, studied medicine and made his way to the Far East and thence to Burma. He served as a medical officer with the British forces in Burma and then embarked on a complex life of intrigue in the Punjab and Afghanistan. At different times, and occasionally at the same time, he served the Sikhs, the British, Dost Mohammed of Kabul and Shah Shuja, the ex-king of Kabul. He was a colourful, eccentric and clever man whose mercurial career deserves a biographer if only for the distinction he enjoys of being the only man ever to have unfurled the Stars and Stripes on the Hindu Kush.

†Avitabile's life is best summed up by the inscription on his tomb at Agerola in Italy.

Lieut.-General Paolo di Avitabile
Born October 1791. Died March 28th 1850.
Chevalier of the Legion of Honour. Of the Order of Merit of San Fernando of Naples. Of the Durrani Order of Afghanistan. Grand Cordon of the Sun and of the Two Lions and the Crown of Persia. Of the Auspicious Order of the Punjab. Naples First Lieutenant. Persia Colonel. France and the Punjab, General and Governor of Peshawar.
A man of matchless honour and glory.
Of all the Europeans in Ranjit Singh's service he was probably the most ruthless. He was also the most successful in that he alone not only amassed a large fortune but managed to get it to Europe and there to enjoy the fruits of his labours.

Ranjit Singh enquired about his travels and Wolff as usual took this as a cue to 'proclaim Christ'. The Maharaja was not even a very good Sikh; such talk bored him. He tried politics but they clearly bored Wolff and by way of distraction the dancing girls were called on. Wolff immediately objected; he 'could find no pleasure in such amusements'. The audience was turning sour. Ranjit Singh tried again and pressed Wolff to a glass of brandy. The Sikh's only tipple was concocted for him by a German doctor called Honigberger from a mixture of raw spirit, crushed pearls, musk, opium, gravy and spices. Burnes and Gerard, Scotsmen hardened on home-made whisky, were the only visitors who were able to stomach it. Later the pale lips of Emily Eden were to be severely burnt by the first sip. Wolff, finding his mouth filled with fire, was unable to swallow.

As a final gambit the Maharaja tried to pull Wolff's leg. Did he believe that nobody could die without the will of God? He did. Then why had Wolff cried out with terror when he was ferried across the Indus? This was a sore point. Few travellers were as physically handicapped as the poor Jew. Woefully short-sighted and a dismal horseman, he also had a horror of water. There had been the shipwreck and then the brush with pirates in the Aegean and before that a lucky escape in the Black Sea. On that occasion he had requested a passage on the SS *Little* from the ship's skipper, Captain Little. But Captain Little had replied that he objected to having preachers on board and anyway his ship was too little to accommodate one. Wolff had found another vessel; the *Little* had been lost with all hands. Crossing the Oxus on the way to Bukhara he had taken the precaution of being blindfolded but, evidently failing to do the same on the Indus, his hydrophobia had been embarrassing to watch. He could only explain it by admitting that his weakness was reason to pray harder. Before he could develop this theme Ranjit Singh closed the audience and the next day Wolff was free to move on towards Ludhiana and Simla.

Nothing in this disastrous interview had been said about Kashmir. When Wolff had first arrived in the Punjab he had written to the Maharaja of his plans and there then seemed to be every prospect of permission being granted. Now his chances looked hopeless. It was a tragedy. If there was one man amongst all the early explorers who might have managed the classic journey to Kashmir, Ladakh and over the worst mountain barrier of all to Yarkand, it was Wolff. The most unorthodox of travellers he was also, to the native mind, the most convincing. It was transparently obvious that he had no political objectives. He claimed to be a *haji*, a dervish or a fakir and even the bigots of Bukhara had come to appreciate that that precisely was what he was. Moreover these religious travellers, along with a few well-known traders, were the only people who could pass freely through the mountains. Wolff had little appreciation of geography and never bothered with maps. It was another point in his favour. Travellers who enquired about distances and directions or who pored over compasses and thermometers were viewed with great suspicion. They might claim that they were travelling purely out of curiosity but no one ever believed them; it was well known that before Europeans invaded a country they sent out surveyors and scientists to record the lie of the land. Finally Wolff was obviously a little mad and this too was an advantage. Madness was thought close to godliness; it was the only qualification possessed by many fakirs and to molest such a man was fraught with dire consequences. At Doab and elsewhere Wolff probably owed his life to this widespread superstition.

For Ranjit Singh's reluctance to let him go to Kashmir, Wolff had only himself to blame. The fatal mistake he made on the way from Wazirabad to Lahore was to issue one of his famous proclamations about 'the personal rule of Christ'. His religious platform at the time had three planks, the conversion of the Jews, the discovery of the lost tribes and the Second Coming. It was the last which caused so much trouble. Having carefully combed through the scriptures, having made complicated arithmetical calculations and having observed the frequency of earthquakes, cholera epidemics and other pestilences, he was convinced that Christ would come again in 1847. If some parts of scripture were interpreted literally then all should be. He had no difficulty believing in miracles. He had heard of thousands, seen some himself and one he had actually worked, the mere casting out of a devil. Visions, too, he took for granted. Besides seeing Christ in Bukhara he had been visited by Saint Paul and the Heavenly Hosts in Malta. Saint Paul, himself a converted Jew and the first great missionary, told his protégé, 'Thou also shalt have a crown but not such a glorious one as I have.'

So why not the Second Coming? The Society for Promoting Christianity among the Jews had not liked it, and for this they had demanded his return to London. Alexander Burnes had also taken exception. He told the Calcutta newspapers that Wolff was preaching the end of the world and that soon everyone would be enjoying the rule of the New Jerusalem and walking about naked like Wolff, with nothing to eat but vegetables. When Ranjit Singh received his copy of the glad tidings he sent it off by express courier to Lord William Bentinck. Bentinck replied asking that the traveller be sent forthwith to Simla. Perhaps the journey had been too much for him; if he had now entirely taken leave of his senses it was vital to get him out of circulation.

Wolff at the time knew nothing of this but flattered by the Governor-General's anxiety to see him he pressed on for Simla. There Lord William, and more especially Lady William who was a devout Christian, decided there was nothing wrong with him. In fact they found him charming. His guileless honesty was irresistible and even about his own vanities and eccentricities he could be disarmingly candid; he was most likely to agree with anyone who called him crazy. When he realised that it was not Ranjit Singh but Bentinck who had scotched his journey through the mountains, he immediately reopened the question. Bentinck tried hard to dissuade him, but when this failed it was agreed that Lady William should make the request of Ranjit Singh. This made it less official and, if anything, more cogent. There followed a long delay, but eventually the Maharaja agreed and Wolff crossed the Sutlej at Bilaspur on September 16th 1832.

From the Governor-General's farewell letter to Wolff it is clear that he was intending not just to visit the valley but to pass through it, heading 'north of the Himalayas' for 'the countries east of Russia'. Like Moorcroft he was trying to reach the oasis cities of Chinese (Eastern) Turkestan, Yarkand, Kotan and Kashgar. Tibet, which he frequently mentions, is not Lhasa but Ladakh, which he would have to traverse on the way. Whether he intended to head from there to China or back to Europe via Russia is not known, but Lord William obviously did not expect to see him back in India. He did not realise that because of Wolff's stupidity over the fatal proclamation it was far too late in the year to reach the high passes beyond Kashmir before the winter.

It is unnecessary to follow Wolff's trip to Kashmir in detail. His route was the same as that used by Moorcroft and Jacquemont on their return from the valley. The important point is that he arrived in Srinagar on October 11th. And left, heading back to India, on the 16th. In other words what should have been an epic journey of

Himalayan travel fizzled out in a brief and pointless excursion. He made one of his rare attempts at descriptive writing when he tried to depict the famous gardens and canals of Srinagar, but failed dismally. The whole place he found a bitter disappointment. 'Instead of the splendid palaces described so enchantingly by the poets one sees only ruined and miserable cottages; instead of the far famed beauties of Kashmir one meets with the most ugly, half starved, blind and dirty looking females.' Four days was scarcely time enough to get even a general impression, but this was obviously not going to be the site of the New Jerusalem.

The only explanation given for his change of plan comes when he is writing of Little Tibet (Baltistan). 'It was much my wish to have gone there, but the snow prevented me and obliged me to return to India.' In so far as it was late to be starting the long journey to Yarkand this makes sense, but the passes into Baltistan would not normally have been closed by mid-October. Wolff is so vague in his geography that he may well have confused Little Tibet with Tibet (Ladakh); certainly the only route to Yarkand of which he knew was the one via Ladakh. On the other hand there were plenty of other reasons for heading south. From Srinagar it is impossible to see the passes to the north or to gauge their condition. Perhaps the Sikhs, not trusting even Wolff to stay aloof from political intrigues on their frontier, preferred to mislead him. Perhaps he was so alarmed by the descriptions of the Karakoram route, which he heard from Yarkandi pilgrims in Srinagar, that he thought better of it. Or perhaps, having briefly retasted the pleasures of civilisation in Simla, he realised he was not ready to return to the wilds so soon. His idea of missionary work was always more like a whistle-stop electioneering campaign than the long hard slog of those who lived amongst their flock. And even if he had wanted to, he could not see the winter out in Kashmir; the Maharaja's permit was good for only a month. With something like relief he ascertained that there were no Jews in Yarkand and quickly turned back before the passes of the Pir Panjal received their first snowfall. He followed the same route over the mountains and crossed the Punjab for Ludhiana and Delhi. This took him through Lahore. Significantly he was not given another audience by the Maharaja.

Wolff went on to make an extensive tour of India before returning to the Middle East and then, by the unusual route of Abyssinia, Bombay, St. Helena and New York, he reached London in 1838.

He made another epic journey to Bukhara in 1844, and he died, the vicar of Ile Brewers in Somerset, in 1862. In his late fifties he was still rueing that lost opportunity in 1832. He wrote to the Royal Geographical Society with a proposed itinerary: 'Bombay, Kashmir, Lhasa, Kashgar, Khokand, Kotan, Yarkand, Ili, Kamchatka, Rocky Mountains, New York, 3 Waterloo Place London and back to Ile Brewers.' It was every bit as ambitious and haphazard as usual and yet, for Wolff in his younger days, not beyond the realms of possibility.

*   *   *

It has to be admitted that the contributions of Jacquemont and Wolff to the geographical understanding of the Western Himalayas were minimal. Compared to Moorcroft's puzzling and painstaking odyssey, their travels have only light relief to recommend them. They were travellers and not explorers, and they wrote as much about themselves as about the lands they visited. If occasionally their observations are subordinated to the demands of good narrative one is prepared to forgive. They were both such extraordinary characters.

Yet it would be wrong, because of this, to regard them as irrelevant. They were public figures in a way that Moorcroft never was and their accounts were assured of a fame out of all proportion to their geographical significance. Before he died Jacquemont had already been officially recognised by being appointed an officer of the Legion of Honour. No time was lost in publishing his letters, and there was even an English edition by 1834. Other editions, some of them pirated, appeared in quick succession. They were immensely readable and did as much as *Lalla Rookh* to popularise the Himalayas.

Wolff's *Researches and Missionary Labours 1831–34* appeared a year later. The map which accompanied them was a disgrace. It showed the Indus rising near Kashgar, the Ganges flowing through Ladakh and a peculiar configuration of mountains; MacCartney had done better twenty years before. Yet the book ran to a second edition in the first year. Wolff was now a celebrity and when, late in life, he retold the story in his *Travels and Adventures*, a reviewer observed that 'there is scarcely a district in the country where the name of Joseph Wolff does not wake smiles and recollections, sometimes ludicrous, sometimes affectionate'. However badly depicted, the mountains beyond Kashmir were known to have defeated the indefatigable Jew. This was inspiration enough for would-be travellers.

It must also be remembered that at this time Moorcroft's journals were still mouldering in Fraser's tin trunks or stacked on Wilson's busy desk. Their very existence was known only to a handful of men. Wolff and Jacquemont thus enjoyed an open field. Excluding Foster's, theirs were the first accounts of Kashmir in over a hundred years. The excursion to the famous valley was rightly regarded as the supreme adventure of each. And if they were not explorers they were certainly pioneers.

Jacquemont, particularly, had also established something of a precedent at Lahore. He had shown that even without a letter of introduction from the British authorities it was possible to win Ranjit Singh's approval for a visit to Kashmir. The chief obstacle appeared to be not the Sikhs but the British. It was true that in 1831 Ranjit Singh had refused a request from Burnes, but Burnes's objectives were so palpably political that this was hardly surprising. Fraser's case was more typical. Despite his qualifications the Hon. Company would not consider his going even to Lahore. As long as the Maharaja lived and his empire prospered, the British were too wary of his susceptibilities to allow their own officers a free hand. The only people with a chance of reaching the mountains were likely to be independent travellers of unimpeachable reputation.

Fraser's case was a particularly sad one because of all the British representatives strung along the north-western boundary, men like Kennedy and the Gerards at Simla or Wade at Ludhiana, Fraser seems to have shown the deepest interest in the Western Himalayas. He had been to Kabul with Elphinstone, he had been in correspondence with Moorcroft and he was a host and a friend to both Jacquemont and Wolff. A strange man of satanic appearance, his character comes through strongly in the accounts of the last two. He liked to keep apart from the other British residents in Delhi. He dressed in native style, spent hours in conversation with venerable Indians and supported a notable harem outside the city. His understanding of caste and custom was profound and his children, of which there were many, were all carefully brought up in the religion and profession of their mothers' families. Yet he was also a knowledgeable administrator and a man of action. He loved to fight and would spring to arms or chase after tigers—he had killed over three hundred—whenever opportunity offered. Perhaps Jacquemont over-romanticises him, but even Wolff realised that, though a religious sceptic, he had depths of character beyond his understanding.

And of course he was the man who had the missing journals of Moorcroft. One cannot help returning to the question of why he was holding on to them. By being the first into print Jacquemont and Wolff were able to steal some of Moorcroft's thunder, but their conjectures about their predecessor were nothing if not tantalising. Probably most of Jacquemont's information came from Fraser. With Wolff this was not the case. He had never heard of Moorcroft until he reached Persia on the way out. He was the first to penetrate the region which had been Moorcroft's undoing and he found both Bukhara and northern Afghanistan still buzzing with rumour and recrimination over the affair. And in the end he was convinced that all three men, Moorcroft, Trebeck and Guthrie, had been murdered. It would have been fascinating to have eavesdropped on the conversations he had with Fraser. Fraser by then had more clues as to what had actually happened than any other man. Perhaps he was planning to write a summary and for this retained the missing journals. We shall never know. Within two years he too was dead, murdered in circumstances as obscure as those surrounding the fate of Moorcroft.

5. Three Travellers

Legend has it that the valley of Kashmir was once an inland sea lapping between the upper slopes of the Pir Panjal and the Great Himalaya. Moorcroft and Jacquemont found fossilised shells which supported this theory and geologists have since confirmed it; the water escaped by gradual erosion at the rock-bound outlet of the Jhelum. Today's traveller who first sees Kashmir from the air will, if it is spring, find the country still inundated. The lakes, left behind when the sea receded, are then indistinguishable from the flooded paddy fields. One looks down on a watery expanse where the roads are all causeways and the villages islands. Later the fields are drained, the rice is cut and the soil becomes baked and brown. Reflecting a soft blue sky the celebrated lakes of Kashmir at last stand out in all their glory. They are seen to be flanked by trees, willow, poplar, chenar* and walnut. and connected by busy canals and the graceful wanderings of the Jhelum. The acres of bright pasture turn out to be beds of lotus and water lily, swards of duckweed and floating fields of melon and cucumber.

Of these lakes the best known is the Dal on the outskirts of Srinagar. It is a fine expanse of water, possibly the loveliest in the world, and rich with romantic associations. On its shores the Moghul emperors built elaborate gardens and from the heat of the Indian summer retired to their shady lawns and graceful pavilions to enjoy a delicate paradise of poetry and love-making. For greater privacy they also constructed islands, one of which at the northern extremity of the lake, is called the Isle of Chenars. Jacquemont captured some of its departed charm when he spent a languid thirtieth birthday there musing on his chances of winning a Kashmir beauty and cooling his ardour by a plunge in the lake. Subsequent visitors were less respectful of its associations, and about 1836 a black marble stone commemorating the first Europeans to visit the valley was erected on the island. Ranjit Singh himself had given permission for the memorial but, to the Sikh governors who also patronised the place, it

* The chenar is the giant Asian plane tree, *Platanus orientalis*, for which Kashmir is rightly famous.

must have seemed like an unnecessary reminder of European interference. Doubtless it was they who turfed it into the lake, for by 1850 it was gone. No European ever saw it and we only know of the inscription from the three travellers who erected it.

Had it survived it would have been one of the very rare reminders of the Himalayan explorers. Leh has no Moorcroft Hotel and Srinagar no Jacquemont Boulevard. Younghusband Glacier and Vigne Glacier might mean something to the Karakoram mountaineer, but more prominent features like Mount Godwin-Austen and Lake Victoria have long since disappeared from authoritative maps. Only in the realms of natural history are the great names remembered—*Gentiana moorcroftiana, Pinus gerardiana, Parnassia jacquemontii, Messapia shawii* and dozens more. Otherwise there are a very few graves, their headstones overgrown and scarcely legible, and that is all. Unlike the other theatres of nineteenth century exploration in Australia and Africa, the Himalayas never saw a generation of colonists who, anxious to boost their right to be there, would preserve the names of the pioneers.

The inscription on the marble slab on the Isle of Chenars read as follows:

Three Travellers
Baron Carl von Hugel, from Jammu
John Henderson, from Ladakh
Godfrey Thomas Vigne, from Iskardo
who met in Srinagar on 18th November
1835
Have caused the names of those European travellers who had
previously visited the Vale of Kashmir to be hereunder engraved
Bernier 1663
Forster 1786
Moorcroft, Trebeck and Guthrie 1823
Jacquemont 1831
Wolff 1832
Of these only three lived to return to their native land.*

The first to arrive in Srinagar on that November 18th was the Austrian, von Hugel. Three years earlier, on his way to the East Indies, he had met Jacquemont in Poona and from him got the idea

---

* This list is far from exhaustive. It should also include Desideri, Csoma de Koros, Lyons and no doubt others.

of visiting Kashmir on his return to India. Jacquemont didn't think much of this 'self-styled' baron. He was reputed to be a great naturalist but was floored when the Frenchman quizzed him about corvine anatomy. Actually the Baron's title was genuine, his connections—including a close friendship with Metternich—were even more distinguished than Jacquemont's, and he was a knowledgeable and systematic botanist. He had none of Jacquemont's charm, but his means and reputation were enough to ensure the co-operation of the British authorities.

Arrived in Srinagar, he set up camp in the garden where Vigne was already established, though away for the day sketching and duck shooting. Almost immediately a European was announced. Von Hugel, expecting his neighbour, called him in. There shambled through the door not the trim English sportsman he anticipated but a long skinny figure with a bony nose and matted red beard. His clothes were Tibetan but too dirty and tattered to be picturesque. His face was haggard and red, the skin torn to shreds by wind and cold. The Baron, normally a most courteous man, stared in amazement. 'Who on earth are you?' he demanded. Unabashed, with great dignity and in a strong Scottish accent which rolled the r's the stranger replied, 'You surely must have heard of Dr. Henderson?' It was a fine effort from someone who cannot have spoken a word of English for several months.

The Baron had heard of John Henderson, as indeed had most of Upper India. He was the *bête noir* of the East India Company even before he disappeared between Ludhiana and Calcutta earlier in the year. Unfortunately no record of his indiscretions has survived. Von Hugel just says that he was such an inveterate critic of the government that he was banned all access to the press. Vigne, who invariably saw only the good in men, calls him a founder of the Agra Bank and 'prominent for enterprising speculation'. He must also have had a yen for enterprising travel for, knowing that as a servant of the Honourable Company he would be severely censured, he slipped across the Sutlej and headed for Ladakh. His destination, apparently, was the source of the Indus. Kennedy at Simla was the first to sound the alarm but by then, following Moorcroft's route through Kulu, he was already in Leh.

It was the year that Ladakh was finally conquered by the Sikhs. Henderson arrived just before their army and the Ladakhis, easily penetrating his disguise as a fakir, assumed he was a British representative come in the nick of time to ratify Moorcroft's 'treaty' of fourteen years before. He was shown how the good vet's garden was still being tended and how his flocks had grown in the loyal care of the Ladakhis. But Henderson knew little of Moorcroft and nothing of any treaty. He was not a British agent, anything but, and was powerless against the invader. Dismayed, the Ladakhis put him under house arrest and hoped that the mere presence of an Englishman would be enough to stay the Sikhs. Henderson escaped. Without money or food he begged his way down the Indus to Baltistan. He lost his horses, his baggage and his two servants but the Baltis, who had just paid a fond farewell to Vigne, helped him across the Great Himalaya to Kashmir, and thus he arrived in Srinagar.

Von Hugel and Vigne, back from his duck shoot, listened to this story with interest. They warned Henderson that in British India there was a warrant out for his arrest and they clubbed together to enable him to continue his travels. His needs were simple. While he stayed with them in Kashmir he slept on the floor rolled up in blankets. The blankets, von Hugel notes, had afterwards to be destroyed; Henderson had long since foresworn soap and clean clothes. To food, too, he was entirely indifferent. The menu for their celebration dinner was 'hare soup, fresh salmon, roasted partridges, and a ham from the wild boar of the Himalaya'.* Henderson was unimpressed. But spying a jar of hot Indian chutney he emptied the lot over the salmon, exclaiming that the greatest of all his misfortunes had been when his 'chatni' reserve was exhausted.

He stayed only a few days in Srinagar and then set off down the Jhelum heading for Balkh in Afghanistan. Eight weeks later he reappeared in Lahore. Again his hosts were Vigne and von Hugel though it was nursing, not hospitality, that he needed. He was a dying man.

*   *   *

But first, 'Godfrey Thomas Vigne from Iskardo', a most important but neglected explorer whose travels take the penetration of the Western Himalayas another long stride towards Central Asia. Presumably because of his Huguenot surname, Vigne has often been taken for a Frenchman. 'Monsieur Vigne', even 'de Vigne', occur in later references to him. In fact the name was pronounced 'vine' and he

---

* Fresh salmon must mean either Mahseer or Himalayan trout. No salmon has ever reached Kashmir and brown trout were not introduced till later in the century.

was as English as could be. His family were wealthy London merchants supplying gunpowder to the East India Company and he was born in the City of London within a stone's throw of the Bank of England. He was educated at Harrow where his name is still scratched on a classroom wall and was called to the bar from Lincoln's Inn. Then in 1830, perhaps bored or just idle, he gave up the law. The family maintained it was because he was 'too delicate', but even in those days the law was not that rough a profession nor was Godfrey Thomas that feeble. He played cricket for the MCC and was an ardent sportsman.

First he paid a short visit to America and then, in 1832, set out for what was to be an equally short tour of Persia. At the end of it, instead of returning home, he took ship from Bushire to Bombay and arrived in India on New Year's Day 1834.* A desert chase after wild asses near Isfahan had laid him low with fever and perhaps he hoped for better medical advice at a large British base. There was something even more casual about his next move. 'As the climate and gaieties of Bombay did not tend much to the improvement of my health I determined, not having had the slightest intention of the kind when I left England, to run up at once to the cool air of the Himalaya.' A run up to the Himalaya was all very well, but there were hill retreats nearer than Simla. One looks for a better reason for five years of continual travel over the most gruelling terrain in the world.

But with nothing more by way of explanation Vigne set off across India. He celebrated his arrival at Simla by trying to cross the 16,000-foot Boorendo pass into Tibet, then followed the mountains round to Mussoorie and adjourned to the plains for the winter. It was something of a royal progress. He had friends in most places and, where he didn't, he soon made them. Few of the famous names in the Upper India of the 1830s go unnoticed. He was delighted with everything and between praise for Major Everest, the Surveyor-General then measuring his baseline for a survey of the central Himalayas, and for Salonica, the grey arab who won the big race at Aligarh, he lets slip a throw-away line; his permit for Kashmir has arrived. It's as if a visit to the forbidden valley were the most natural thing in the world and the Company's permission the merest of formalities. Nothing could be further from the truth, of course, but this off-hand attitude had its advantages. It relieved him of the need to offer

* Vigne actually says 1833 but this must be a slip of the pen since at the beginning of 1833 he was at Trebizond en route to Persia.

explanations on such delicate points as why he wanted to go there, why the Company agreed or why Wade from Ludhiana made special representations on his behalf to Ranjit Singh. In June 1835, complete with guns, rods and easel and sporting his usual outfit of Norfolk jacket and broad-brimmed hat, he straddled a buffalo skin at Bilaspur, crossed the Sutlej and headed for Kashmir.

Inflated buffalo skins

Impulsive, entertaining and unbelievably casual G. T. Vigne is nevertheless a recognisable figure. After the enthusiasms of Moorcroft, the affectations of Jacquemont and the ravings of Wolff one shakes his outstretched hand with a sigh of relief. Here at last is a reasonable man. He takes things as they come; pragmatic common sense is his speciality. It is easy to identify with Vigne, and it would have been pleasant to travel with him. His charm is neither florid nor demanding but a quiet and genial affability. There is even a mild sense of humour, something almost unique amongst nineteenth century explorers. Away from the Himalayas he would have been happiest alone, for he was an independent character, with his gun and his dogs looking for wigeon along the Thames estuary. Surely no one who has read his books could imagine he was other than English.

Hunting to such a man was more than just a pastime. His brother Henry, who was also deemed too delicate for the bar, rejoiced in the nickname of 'Nimrod', 'mighty hunter before the Lord'. Godfrey Thomas, no less a shot and never one to be outdone, was generally known as 'Ramrod'. On his way to the Himalayas he had shot pheasant, partridge, snipe, quail, alligator, antelope, tiger, panther and 'mullet as they rose to the surface'. He had coursed with cheetahs, with hounds and with falcons and with the rod in the Elburz had 'killed six or seven dozen trout a day'. That he could find no hares in Kashmir was something that worried him deeply. It was such perfect country. On the other hand, when his terrier flushed out a fox which was 'not the little grey leading article of Hindustan but the large full brushed Meltonian', he was markedly reassured. Sport must have inspired his first interest in the mountains and he now heads the distinguished role of Kashmir sportsmen, which includes African travellers like Speke and Grant as well as most of the Himalayan pioneers.

More surprisingly, Ramrod Vigne was also a good amateur botanist and geologist and an accomplished artist. When Ranjit Singh asked what he did he replied, rather curtly for a gentleman is not usually asked such a question, 'I can draw'. He certainly could and his picture of the Maharaja looking like a satanic cyclops is probably no harsher than the reality. In 1842 he exhibited at the Royal Academy and in spite of a prolific output his work, particularly the arresting portraits, still commands a good price. His wide-angle panoramas were less successful. The eye is taken by the foreground detail, a group of figures or a dog always of sporting potential, while the faithful perspectives of the whole somehow fail to capture the peculiar character of the country; to any but a resident his view of Kabul could pass for Srinagar and vice versa.

This failure was more than retrieved by his fine prose descriptions. With Vigne we at last get a vivid impression of 'the untried and fairy wilds of Kashmir'. Over five years he came to know the valley better even than had Moorcroft. He loved it dearly and finding no fault with Tom Moore's licence declared it 'the noblest valley in the world'. He saw not just the lakes and the gardens, the trees and the mountains but the whole context of the place, something which the visitor feels but rarely analyses.

Innumerable villages were scattered over the plains in every direction, distinguishable in the extreme distance by the trees that surrounded them; all was soft and verdant even up to the snow on the mountain top; and I gazed in surprise, excited by the vast extent and admirably defined limits of the valley and the almost perfect proportions of height to distance, by which its scenery

seemed to be characterised. . . . Softness mantling over the
sublime—snugness generally elsewhere incompatible with extent
—are the prevailing characteristics of the scenery of Kashmir;
and verdure and the forest appear to have deserted the countries
on the northward, in order to embellish the slopes from its snowy
mountains, give additional richness to its plains and combine with
its delightful climate to render it not unworthy of the rhyming
epithets applied to it in the east of:

Kashmir bi nuzir—without equal
Kashmir junat puzir—equal to paradise.

Though much the largest of the Himalayan valleys it is perhaps the
least dramatic. There is none of the savage majesty of Lahul, the
Alpine clarity of Kulu or the exotic contrasts of Chitral. Its beauty is
more mature and stately. It lies precisely in those 'perfect proportions of height to distance' and the 'softness mantling over the
sublime'.

Vigne's descriptions of Kashmir have been raided by many a later
travel writer. The country has changed little and they are now as apt
as ever. The city of Srinagar is still 'an innumerable assemblage of
gable-ended houses, interspersed with the pointed and metallic tops
of masjids or mosques, melon grounds, sedgy inlets from the lakes
and narrow canals'. A teeming but quite un-Indian city, it is both
fascinating and slightly frightening. There are few streets, but there
must still be many a narrow alley and smelly little canal to which no
European has ever penetrated. The inhabitants too are still 'a lying
and deceitful race of people'. But Vigne, and for this one must like
the man, admitted what very few of his fellow countrymen would
ever appreciate, namely that 'when detected in a fault their excuses
are so very ready and profuse, and often so abound in humour, that it
is impossible to abstain from laughing and to attempt an exhibition
of anger becomes a farce'.

In all, Vigne crossed the Pir Panjal to the south five times, twice
he took the western approach by the Jhelum valley, once he left by an
eastern route to Muru Wurdan and Kishtwar and six times he crossed
the Great Himalaya to the north. He lists twenty passes into the
valley and had personal knowledge of most of them. The great
revelation of his Kashmir travels lay not so much in the glories of the
Dal Lake or of the Vale itself but of the surrounding mountains and
valleys. Later he was criticised for the confused presentation of his

6 Godfrey Thomas Vigne, the Skardu Valley
in the background

7 Srinagar: the Jhelum River. From a line drawing
by G. T. Vigne

8 The Indus in Baltistan. From a line drawing
by G. T. Vigne

material and his vagueness about dates and itineraries. It is extremely difficult to piece together his precise movements year by year. But one sympathises with his problem. Maybe he was not particularly methodical, but to have presented his travels in journal form would have been an impossible task. He covered so much ground and so often retraced his steps or recrossed a previous route that even a map of them would present a meaningless tangle.

Instead he wrote what is really a guide book and one which thirty years later was still the best of its kind. His wanderings south of the Great Himalaya extended from Chamba on the confines of Kulu to Muzafarabad, a good two hundred miles to the west. Only half of this sweep is occupied by the vale of Kashmir. The rest is extremely mountainous, but alpine rather than Tibetan, a land of fast running rocky rivers in steep gorges with fine forests above giving way to birch, rhododendron, juniper scrub and finally, between the tree-line and the snows, rich mountain grazing. Much of it was new to geography; he was the first European to plot the courses of the Ravi, Chenab and Jhelum rivers through the Pir Panjal and the first to visit Kishtwar, Chamba, Bhadarwah and Muzafarabad.

But to Vigne geography was not everything. Moorcroft and Jacquemont had both been appalled at the misgovernment of Kashmir and had reported that the extension of British rule was already earnestly desired by most of the people. Trebeck put it more forcibly. Annexation would no more constitute unwarranted interference than 'seizing the hand of a suicide or arresting the blow of a murderer'. 'And I wish to Heaven, Leeson,' he had written to his Irish friend, 'that some thousands of your own distressed countrymen and of our emigrants had a footing in the heart of Asia.' Vigne took up the cry and it was repeated by many subsequent visitors. Kashmir was not India but could and should be, as it was under the Moghuls, the brightest jewel in the imperial crown. Psychologically, strategically and economically it represented, and still does, an invaluable acquisition.

For the British it had the added attraction of a European climate. With the annexation of Kashmir, Simla would be promptly deserted and the valley would become the most popular retreat in the East. The Company's troops would have the world's finest and healthiest training ground and their officers, instead of succumbing to the heat and the bottle, would be hardened in pursuit of ibex high above the snowline. British capital and know-how combined with the skills of the Kashmiri craftsman and farmer would make the land as prosperous as it was beautiful. It would become 'the focus of Asiatic civilisation; a miniature England in the heart of Asia'. Vigne could scarcely look on the famous ruins of Martand without noting of the level ground beside them 'a nobler racecourse I have never seen'. He failed to foresee Gulmarg, then just an open saucer of turf fringed with pines and glaciers, becoming a golf course and ski resort, but he reckoned that with the addition of a herd of deer and a mansion it would make a fine English park.

It was only twelve years since Moorcroft's visit but Vigne, born in 1801, was of a totally different generation and background. Moorcroft had filled his journals with notes on how the skills and produce of the Himalayan lands could be adopted in Britain and India. Vigne conjectured on what the introduction of British skills could make of the Himalayan lands. Fresh from America and with none of the commercial priorities of a servant of the East India Company, he was thinking less of 'improvements' and trade and more of colonisation. Like Trebeck he dreamt of a little oasis of English life in the heart of Asia, a place where the rulers of India, having left the heat in the plains and having dumped the White Man's Burden on the Pir Panjal, could change into their old tweed suits and settle down before a good log fire on their own few acres.

Vigne is not as specific as this, but one can see how his mind was working. The memorial on the Isle of Chenars was meant for a future generation of English settlers in Kashmir. Three quarters of his *Travels in Kashmir, Ladakh, Iskardo etc.* is devoted solely to the valley. Thanks to him it did eventually attract a flood of Englishmen and indirectly came under British protection. But it never became a colony; in fact it was never a part of the British Raj and technically remained foreign soil to British and Indian alike until 1948.

Tucked away at the end of Vigne's work on Kashmir lies the account of his wanderings north of the Great Himalaya. It is almost an appendix and unlike the rest of the book is reticent as well as discursive. He well knew the geographical interest that would attach to this part of his travels. Some of his letters from Baltistan, waterlogged and scarcely legible, were published as soon as they reached Calcutta. The Royal Geographical Society was given an outline of his journeys directly he returned and Sir Alexander Burnes, the then authority on Asiatic travel, wrote of his anxiety to know more. But the book when it appeared in 1842 begged far too many questions. Why, when it was Kashmir that he went to see, did he make straight for Baltistan? And return there again and again? Why were the British authorities so unusually co-operative and why so Ranjit Singh, whom he did nothing to humour? And where was he really heading as he probed the mountains from east to west? One plots, rather than follows, the course of his journeyings and all the time one wonders whether it was just the man's natural vagueness masking an insatiable curiosity or whether he really had something to hide.

The route from Kashmir to Baltistan first takes the traveller over the Great Himalaya by two passes, the Gurais and the Burzil of about 12,000 and 13,000 feet respectively. This is not particularly high by Himalayan standards but, like the Rohtang and the Zoji, they attract a colossal snowfall and are impracticable for at least six months of the year. Vigne first reached Srinagar in early August 1835, and anxious to avoid being caught by a freak snowfall like that which had reputedly checked Wolff, pressed on for the mountain wall above Bandipur. On August 29th he camped below the Gurais pass. This was as far north as Jacquemont had got; from now on he was breaking new ground. The first surprise came at the top of the pass when the Balti agent sent to escort him suddenly galloped off. He was shouting something about a priceless excitement.

> I quickly followed him and the stupendous peak of Diarmul or Nanga Parbat, more than forty miles distant in a straight line, but appearing to be much nearer, burst upon my sight, rising far above every other around it and entirely cased in snow excepting where its scarps were too precipitous for it to remain upon them. It was partially encircled by a broad belt of cloud and its finely pointed summit glistening in the full blaze of the morning sun, relieved by the clear blue sky beyond it, presented on account of its isolated situation an appearance of extreme altitude, equalled by few of the Himalayan range, though their actual height be greater.

Actually the appearance of extreme altitude was not an illusion. Nanga Parbat is not '18,000 or 19,000 feet' but 26,500. Vigne, ever modest with his altitudes, was, like the rest of his generation, as yet unaware that the peaks of the Western Himalayas were as high if not higher than those further east. Nanga Parbat was at the time the second highest mountain known to geography. It is not as ethereal as Kanchenjunga seen from Darjeeling nor as staggering as Rakaposhi cleaving the heavens above Hunza. It is too massive, too ponderous to be elegant. Yet it must have been a stunning discovery, this juggernaut of a mountain dwarfing the whole jagged horizon. It marks the western extremity of the Great Himalaya and lies almost at the dead centre of the mountain knot which is the Western Himalayas, a fitting climax to both.

At the village of Gurais, Vigne stocked up with provisions for the hundred miles of uninhabited wilderness that lay ahead and on September 1st with forty-five porters set off for the Burzil. With his own servants and those of the Balti agent, Nazim Khan, the party numbered about sixty. Von Hugel later complained that Vigne was far too indulgent with his men and that his personal servant, an Anglo-Indian called Mitchell, was a confirmed drunkard. Vigne certainly provided a tent for his men, a consideration not shown by the likes of Jacquemont, and was generally a sympathetic master. He was quite happy with Indian food and his only eccentricity seems to have been a concern that at any given moment on the march he should be no further from a cup of tea than the time it took to boil a kettle.

Normally the porters and kitchen servants would leave camp first and make straight for the next campsite. Vigne, having struggled into his boots and stuffed his slippers and a dry pair of socks into his pocket, would then follow by a more devious and interesting route along with his own little flying column. First came a chair, 'the indispensable accompaniment of dignity', in which were strapped thermometer, sextant, sketching materials and the vital tea-making equipment. The man underneath it was the guide. Next, puffing and blowing, came a man decked about like a Christmas tree. This was the thirsty Mitchell. In one hand he carried an umbrella and in the other a second thermometer. From his belt dangled a geological hammer, across his shoulders was slung a telescope and behind his back, wrapped in a cloth and secured goodness knows how, was a weighty plant book. Then followed Ramrod Vigne, gun in hand if there was a chance of partridge and on foot for the ascent of the Burzil. With him would be his *munshi* (secretary and interpreter), a man burdened more by responsibilities than luggage. And finally the groom, a wiry little man tugging frantically at a wiry little horse.

Four days out from Gurais they were encamped just below the pass and just above the treeline when, in the fading light, Vigne spotted a Balti scout on the rocky skyline. High passes have an eerie stillness about them. The traveller, feeling exposed and slightly edgy, stays close to the camp fire and the warmth of companionship. Suddenly out of the brooding silence came a noise like no other on earth, 'the loud, distant and discordant blasts of Tibetan music . . . the sound grew louder and louder and we were all on the tiptoe of expectation'. From the darkness emerged a band of fifes, clarinets and six-foot-long brazen trumpets. Forty soldiers, 'the wildest looking figures imaginable', with their hair hanging down in long ringlets on each side of the face, followed and with them their young commander, a son of Ahmed Shah. The Raja, it transpired, was only a few miles ahead. He had issued forth from Skardu to intercept a band of robbers and was now lying in ambush for them just over the pass. Vigne would meet him next day as soon as the action was over.

A lot depended on this meeting. Vigne was well prepared. As presents he had knives, shot-belts and powder horns, pistols and telescopes. By way of diversion there was a folio of portraits and pin-ups, some prints of racehorses and hounds and even 'portable machinery for chemical experiments'. But what he didn't have, and what Ahmed Shah was bound to look for, was any diplomatic status. Moorcroft, Jacquemont and even Wolff had all been approached by emissaries from Baltistan. In spite of Jacquemont's insistence that the Sikhs would never bother such a poor and isolated kingdom, they had the following year invaded it. Ahmed Shah had seen the fall of Kashmir and of Ladakh. In 1833 he had managed to hold them off, but he knew that they would be back and that his only hope of immunity lay in a British alliance. At last he had made contact with Wade at Ludhiana, and to him officially had renewed his request that a British officer be sent to Skardu.

This much Vigne must have already been told by Wade. Rightly, it would seem, he interpreted it as a hint that the Company, although still unwilling to send one of their own people, would support an attempt by him to reach Baltistan. Wade would manage things with Ranjit Singh and would be interested in his report. But no more. He was to remain a private traveller and would get neither help if he ran into difficulties nor acknowledgement if he returned. Though Ahmed Shah would assume that anyone from British India who reached his kingdom must have official backing, he was to disclaim it completely and to offer nothing.

Such appears to have been Vigne's understanding of the situation. The only flaw in the argument is that whatever Vigne or Wade might say, neither the Sikhs nor Ahmed Shah were likely to believe a word of it. When Wade supported Vigne's hastily written request from Kashmir that Ranjit Singh allow him to proceed at once to Skardu, the Sikhs can have had no further doubts. And Ahmed Shah was pretty sure to encourage their suspicions. The presence of a British Agent at Skardu was his surest guarantee against attack. In fact Vigne had less cause for concern over the welcome he would receive than about whether he would ever be allowed to depart.

There are four elements here: the ambivalence of the British government, the enigmatic status of the traveller, the suspicious attitude of the Kashmir authorities (Sikhs but soon Dogras) and the anxiety of some remote princeling for a British connection. Often conflicting, these were to bedevil attempts to penetrate the Western Himalayas. We have seen them in Moorcroft's story and they will recur in the travels of Johnson, Shaw and Hayward. In the light of what happened to the others, Vigne was extraordinarily successful in avoiding a combination of these factors which would prove explosive. One begins to wonder whether he was quite as casual and unsystematic as he appears. It is seldom possible to penetrate the blanket of secrecy on the British side or to untie the web of intrigue on the native side. We can only judge by how each traveller managed. Vigne had taken a high-handed stance when the Kashmir authorities had tried to impede his departure for Baltistan. He was now, with equal success, about to deploy a gentler approach to Ahmed Shah.

The two men met as soon as the robbers had been successfully annihilated. Vigne in his broad-brimmed white cotton hat and white duck-shooting jacket was approached by a tall and imposing figure who doffed his turban and frequently stopped to salaam deeply. The Englishman took his hand and wrung it vigorously, explaining through the interpreter that such was his native custom. Ahmed Shah was delighted. It was his lifelong ambition to meet a 'feringi' and now it had happened on a day when he had already fought a successful action. So far so good, thought Vigne, and after more mutual flattery he was so emboldened as to ask leave to sketch his host on the field of battle. Given the situation, the result shows a remarkably steady hand.

For the next three days they travelled together across the bleak and treeless Deosai plateau. The altitude was about 12,000 feet, and even now in early September there was a heavy frost at night. The Raja presented Vigne with a pair of warm Tibetan socks. He returned the compliment with a bottle of brandy which was sent on under escort to Skardu. They were getting on famously. The Raja had 'some excellent English ideas about him' and when Vigne exhibited his pin-ups, 'engravings of Chalon's beauties', he gazed in silent admiration. There followed a print of King William IV and his consort, and the Raja insisted on writing a personal note in the margin sending to His Britannic Majesty respectful salaams and an earnest wish for protection. Vigne declared from the start that he had no political status, but he admits that till the very last Ahmed Shah did not appear to believe him. Whether and how this prejudiced their relations it is hard to say. But regardless of it there developed between them a real and lasting friendship the like of which is not to be found between later explorers and explored.

On September 6th Vigne at last got his first glimpse of Baltistan. In the morning the Raja had pointed to a line of mountains along the rim of the Deosai from which they would descend on Skardu and the Indus valley. By sunset they had reached the foot of this ridge. Vigne could contain himself no longer and galloped on ahead. He abandoned his horse on the steep climb and finally stumbled over the upper edge of a glacier and peered into the twilit distance. '. . . through a long sloping vista formed of barren peaks, of savage shapes and various colours, in which the milky whiteness of the gypsum was contrasted with the red tint of those that contained iron—I the first European that had ever beheld them (so I believe), gazed downwards from a height of six or seven thousand feet upon the sandy plains and green orchards of the valley of the Indus at Skardu, with a sense of mingled pride and pleasure, of which no one but a traveller can form a just conception.' He could see the rock of Skardu, the Raja's stronghold, rising like a second Gibraltar out of the level bed of the Indus and beyond it the Shighar valley whose river joins the Indus at the base of the rock. Further north and 'wherever the eye could rove, arose, with surpassing grandeur, a vast assemblage of the enormous summits that compose the Tibetian Himalaya'.

Vigne usually calls the country Little Tibet, and both the people and the terrain are distinctly Tibetan in appearance. The valleys are perhaps deeper and steeper than further east and they are separated not by rolling plateaux but by lofty spurs. Yet there is the same over-all impression of rock and sand, harsh white light and biting dry wind. Natural vegetation is a rare and transitory phenomenon; cultivation just an artificial patchwork of fields suspended from a contour-clinging irrigation duct or huddled on the triangular surface of a fan of alluvial soil washed down from the mountains. The water is an icy grey mercury fresh from the glaciers and glittering with mica. Shade, the other essential, is either the dappled pallor afforded by willow and apricot tree or the deep and shivery gloom of a Balti house.

The people themselves have the narrow eyes and careworn wrinkled faces of the Tibetan. But they are not Buddhists. Long ago they were converted to Shiah Mohammedanism and represent the most easterly outpost of Islam in the Himalayas. In place of the polyandry of Buddhist Tibet, which operates as a form of population control, they adopted the polygamy of Mohammed which does quite the reverse. A Balti's needs are about as basic as any man's could be. He can sustain his habitual cheerfulness and considerable stamina on a diet consisting entirely of raw apricots and a dough made from barley flour. Both, and little else, are grown extensively, but seldom in sufficient quantity to support the ever-growing population. The Balti has been forced to seek a livelihood elsewhere, usually as a porter or coolie. Vigne was the first European to exploit this potential. Later they became the essential workhorses of every expedition towards Gilgit and beyond.

The 'vast assemblage of mountain summits' which he saw from the ridge above Skardu was the next barrier between India and Asia. At the time he called it the Tibetian Himalaya but later, and more correctly, the Mustagh or Karakoram range. For the next few weeks, for five months in 1837 (1836 he spent mostly in Afghanistan) and for a similar period in 1838, he relentlessly probed this mountain barrier. It is hard to say in which year or in what order but four major journeys are discernible. To follow each one in detail would be tedious, but they are so important to the future course of Himalayan exploration that a note of each is essential.

Working from west to east the first took him from Skardu to Astor, a fort commanding the Kashmir-Gilgit road, and thence to a vantage point called Acho overlooking Bunji on the Indus. From here he could see Gilgit, and he intended to cross the river and proceed there. His *munshi*, who had been sent on ahead, gave conflicting reports about his likely reception but Vigne reckoned all would have been well if he had been able to approach without such a large Balti escort. As it was, the Gilgitis took fright. They saw what looked more like an army approaching and promptly destroyed the bridge across the Indus. Later a similar incident nearly put paid to his visit to

Ladakh. One cannot avoid the conclusion that Vigne was a little gullible where the good faith of his Balti hosts was concerned. At that time there was in fact no bridge across the Indus at Bunji; the first was built in the 1890s. There was a ferry and this may well have been destroyed, but it seems more likely that the Baltis were jealous of their guest and simply invented the story to prevent his opening relations with their traditional enemies.

The second journey from Skardu was north up the Shighar and Basho valleys into the heart of the Karakorams. On the crossing to Astor, Vigne had negotiated some large glaciers, experienced snow blindness and made his always conservative estimate of the altitude nearly 16,000 feet. He was a brave and determined traveller, fit and to some extent acclimatised. But this is not adequate preparation for the Karakorams. The 'delicate' barrister was no mountaineer; he had neither the experience nor the equipment. Indeed it is doubtful if anyone at that time could have penetrated, let alone crossed, this appalling wilderness. Vigne tried. He attempted to cross over the Hispar glacier to Hunza, intending to proceed across the Pamirs to Khokand, all of five hundred miles away in what is now Soviet Russia. He also investigated the possibility of crossing from Shighar to Yarkand by what later became known as the Mustagh pass. But both routes were way beyond his capability. The Hispar route was not crossed until Sir Martin Conway, the greatest mountaineer of his day, did it in 1892. The other route defeated Godwin-Austen and, fifty years later, Sir Francis Younghusband only forced his way through by taking the most lunatic risks. Without much prompting this time, Vigne turned back from the Basho valley.

As a result of Moorcroft's try for Yarkand the Chinese, so Vigne was told, had executed a massive wall painting of a European so that every Yarkandi would instantly recognise and apprehend any subsequent visitor. Burnes heard similar stories on his way to Bukhara, only his version had it that anyone who recognised such a *persona non grata* was entitled to appropriate him and his possessions, forwarding to the authorities as evidence just the man's head. Vigne decided it would be best to avoid the Chinese, but having established that routes to the north and west of Skardu were impractical, he turned towards the main trans-Karakoram track from Leh hoping that once across the mountains he would be able to strike west and thus avoid Yarkand.

It was probably in 1837 that he travelled up the Indus to Leh with the professed intention of reaching a glacier lake at the head of the Nubra valley. This, he was told, was the source of the Shyok, a major tributary of the Indus and by some geographers thought to be the Indus itself. On the way he inspected the confluence of the two rivers and rightly pronounced that the one from Leh which Moorcroft had explored was the parent stream. But he was still interested in the Shyok, particularly as its supposed source lay so near the Karakoram pass on the Ladakh–Yarkand route. This was the track that Moorcroft had intended to follow. It was regularly used by merchants and so well within Vigne's capabilities. But he was reckoning without the Sikhs. They were now established in Ladakh and had no intention of letting him befriend any more of their mountain neighbours.

It was bad enough that he had been given a free hand in Baltistan. Vigne in his dealings with native authorities is usually a model of good humoured patience, but the treatment he received in Leh was too much. Under constant surveillance, forbidden contact with the Ladakhis and prevented from organising his onward journey, he levelled bitter accusations of insolence and disrespect. A complaint was lodged with Ranjit Singh and an appeal made to Wade. They were successful but the necessary directives came too late. By the time he got away from Leh winter was imminent. He reached the lower end of the Nubra valley, no further than Moorcroft, before he was driven back.

In 1838—and this date at least seems certain—he made his final attempt, again via Nubra, to reach Central Asia. To avoid another brush with the Sikhs he this time tried to cross direct from Skardu to Nubra, thus bypassing Leh. Such a route existed but for reasons that became obvious it was rarely used. Striking up the Saltoro valley the expedition encountered torrential thunderstorms. The tents must have been so soaked that they were left behind to dry, and the next night they slept in the open. By then they were well up one of the vast Karakoram glaciers. Camping on ice is never pleasant but, without cover, it must have been misery. They managed to brew a pot of tea and, thus fortified, Vigne swathed himself in blankets and stretched out on the hard ice. Again the weather deteriorated. Morning found him soaked through with sleet and buried under snow. Still, and with three more such nights ahead, he wanted to go on. His guides refused; with the glacier's crevasses hidden under new snow they were right. Bitterly disappointed, he turned back for the fourth time.

To dwell on Vigne's failures would be churlish. He had achieved too much. He was the first European to visit Baltistan and Astor and the first, from that vantage point above Bunji, to sort out the complicated topography of Gilgit, Yasin and Chitral to the west and of Chilas and the so-called Indus valley states to the south-west. He was the first to give a recognisable account of Nanga Parbat; and the whole Karakoram system was virtually his discovery. At the time it was widely believed that beyond the Great Himalaya north of Kashmir there stretched a vast plateau which extended down to the deserts of Turkestan. Vigne showed that nothing could be further from reality. It was a region of mountains every bit as high and more impenetrable than the main Himalayan chain. Whether he ever saw, rising from this sea of rock and snow, the distant giants of Masherbrum, Gasherbrum and that shyest of all peaks, K2, we don't know. But in an age when geographers were still doubtful whether glaciers could exist in any latitude warmer than that of the Alps, he brought back accounts of one of the world's greatest glacial systems. He had had no chance of ascertaining their length, but the snout of the Chogo Lungma glacier he described as a wall of ice a quarter of a mile wide and a hundred feet high. From a cavern in the clear green ice there flowed not a stream, 'no incipient brook, but a large and ready formed river'. It roared forth, shunting enormous blocks of ice against its rocky bed with a noise like distant cannon; it was the grandest spectacle that he saw on the whole of his travels.

The discovery of this massive mountain system flanking the northern banks of the Shyok and Indus from Leh to Hunza put a whole new complexion on the Western Himalayas. Geographers began to get some inkling of the depth of the mountainous country north of the Punjab and of the fact that they were dealing not with a continuation of the Great Himalaya but its conjunction with a web of other mighty systems. As for future travellers, they could now forget about reaching Central Asia by striking due north for Skardu. Baltistan was a dead end. Vigne had clearly shown that the only possible routes lay west of the Karakorams via Gilgit or east via Leh. And of these the latter was the most promising. For the first time the ranges and passes to be crossed en route to the Karakoram pass were correctly set out, and on his map Vigne even suggests that there might be a further range beyond the Karakorams.

In November 1838 he left Baltistan for good. Within a year the country had been taken by the Sikhs. Both he and Ahmed Shah, and no doubt Wade too, had foreseen this. For four years his presence there, and that of a Balti representative in Ludhiana, had stayed the onslaught.* Now, for some reason, it was no longer considered necessary to bolster Balti independence. One can only speculate, but had Vigne found a practicable route from Skardu to Central Asia would the British have been so willing to abandon Ahmed Shah? Almost certainly not. And this being the case, it is worth looking again at Vigne's status vis-à-vis the British authorities. In 1836 Wade, their representative at Ludhiana, wrote an interesting letter to the Surveyor-General. He had just received the report of a native whom he had sent to explore the country beyond Kashmir. The exercise had been a success but he still regretted that, since Moorcroft, so little had been done in this direction. To make good this deficiency he recommended that 'the best mode of all would no doubt be to employ an ... enterprising European officer ... without being further accredited than were Mr. Moorcroft, Lieutenants Connolly and Burnes and the late Dr. Henderson'. It seems highly probable, and for a man like Wade it would not have been out of character, if he had already made just such an arrangement with Ramrod Vigne.

But if this most plausible of sportsmen was really some sort of spy, what then was his assignment? It could have been just to woo Ahmed Shah and explore his country. More probably it was to penetrate beyond Baltistan and to assess the country's strategic importance in continental terms. Meddling with the expansion of the Sikh empire mattered less than anticipating the expansion of the Russian empire. Vigne, of course, says nothing about this. But he does talk a lot about routes to Khokand, from which direction a Russian advance would most probably be made. He does speculate on the movements of merchants between Russia and India. And he did report to the Viceroy on the presence of a Russian agent in Shighar.

The British attitude towards Russia had changed significantly since Moorcroft's day, and the late 1830s found India convulsed by one of its worst ever bouts of Russophobia. In 1838 an ill-fated army was despatched into Afghanistan with the sole idea of countering Russian designs there. Two years before, Vigne himself had spent a summer in Afghanistan exploring a little known route into that country and then trying to cross the Hindu Kush to meet

---

* Vigne also acknowledges Henderson's visit to Baltistan in 1835 and that of H. Falconer, Superintendent of the E.I.C.'s botanical garden at Saharanpur in 1838.

Moorcroft's old enemy, Murad Beg of Kunduz. As usual he had travelled in a private capacity but again it is difficult to believe that he would have been permitted to go simply to satisfy his own curiosity. The times were too critical; the man himself too plausible. The Great Game, which came to dictate the whole course of exploration in the Western Himalayas, was already influencing the explorers. Vigne looks like one of its least conspicuous and therefore most successful players.

After fifteen years, the voice of Moorcroft crying out of the wilderness of Ladakh was at last being heeded. There was concern that the Western Himalayas might screen some vulnerable back door into India. Hence the radical departure from previous policy over visitors to Kashmir which Vigne's travels suggest. And hence his returning to Baltistan year after year. It was still the same happy-go-lucky Vigne but he had become involved in something which, for all its excitement, had its drawbacks. He had grown to like the simple Baltis too much, he appreciated their savage scenery too keenly and he esteemed their Raja too highly. It was not in his nature to abandon them to the grasping Sikhs. If their only salvation lay in his finding a flaw in the mountain wall on which to make out a case for a permanent British interest in the region, he was determined to find it. This alone explains his endless probing of the Karakorams. It also explains his intense disgust over the prevarication he encountered at Leh and the appalling risks he was prepared to take on that last desperate bid to force a way up the Saltoro glacier.

When he failed to establish that Baltistan had any strategic value, there was no point in gainsaying the Sikhs any longer. Bentinck was gone, and with him the Company's cautious external policy of the last twenty years. But it was not the Sikhs who were to be the victims of the first new wave of British expansion. They were to be the allies. In 1838 the new Governor-General was suing for Ranjit Singh's co-operation in the invasion of Afghanistan. Part of the price paid for this was recognition of Lahore's rights to Kashmir and Ladakh and a free hand in the rest of the mountains, including Baltistan.

*   *   *

The famous meeting of Vigne, von Hugel and Henderson took place after Vigne's first visit to Baltistan, when he was on his way back to the Punjab. Kashmir in November, as the Baron was for ever observing, is a cold and cheerless place. Henderson soon set off on his last escapade and the other two, joining forces, headed for Lahore. Neither had as yet met the Maharaja though indebted to him for their safe conducts. Courtesy as well as curiosity dictated a visit.

The Baron had spent just three weeks in the valley and had broken no new ground. Undeterred, he later wrote a comprehensive account of the region entitled *Cachemire und das Reich der Siek*. This drew heavily on information gleaned from Vigne and Henderson but curiously ignored Moorcroft's *Travels*, then at last in print. It was rather like writing on the Later Roman Empire without consulting Gibbon. Von Hugel says that he was anxious to preserve the originality of his own impressions but, as a result, his summary of Moorcroft's achievements was absurdly and indefensibly unjust.

His own observations were not much better. He overestimated the altitude of Kashmir by 1,000 feet and denied what every other visitor had found self-evident, namely that the valley had once been a lake. It follows that as a Himalayan explorer von Hugel deserves little space. Yet, and this surpasses understanding, it was to him that the Royal Geographical Society in 1849 presented its Patron's Gold Medal. It was for this particular journey and for the book that resulted from it. Both he and Vigne contributed to the Society's journal and both had their books reviewed in it. The Society had all the facts. Yet von Hugel was honoured and Vigne overlooked. The Baron was taken seriously but Vigne was regarded as unreliable. The latter's descriptions were thought picturesque but fanciful and his maps and measurements were approached with great caution. Not until 1861, by which time there was a framework of surveys against which to measure his achievements, did a contributor to the Society's journal at last make amends. Vigne's book was now acclaimed as the best work on Kashmir and more practical and trustworthy than von Hugel's. His map was 'the mine whence others were manufactured; and when the time and circumstances under which it was compiled are considered it must be regarded as an outstanding production'. Later authorities have upheld this opinion, with Sven Hedin crediting Vigne with a clearer idea of the Karakorams than Shaw entertained thirty years later and Kenneth Mason calling his book 'a classic' and the first comprehensive account of the Western Himalayas.

To Vigne, after a taste of Baltistan, the journey to Lahore must have been child's play. But to von Hugel it was a Calvary, and this in spite of a style of travel undreamt of by his companion. Sometimes the Baron rode, more often he was carried in a sedan chair by twelve bearers; he walked only to stretch his legs. His principal tent had poles twenty-five feet high and the roof alone weighed quarter of a ton. The kitchen was stocked as for a world cruise with 'preserved meats hermetically sealed in tin boxes, wines and drinks of various kinds, preserved fruits and sweetmeats'. As stocks ran out they were replenished by convoys from Ludhiana. And still the Baron was 'worn out by indifferent food'. He felt so ill and exhausted that he cast himself down and thought it madness ever to hope to see his friends again.

He was also desperately lonely. Henderson would have had some grounds for complaining of the solitude of the traveller but not von Hugel. Sixty porters and seven mules carried his luggage while thirty-seven servants ministered to his every need. There was a secretary, an interpreter, a torch-bearer, a butler, three cooks, a water-carrier, a tailor and a man for lighting his pipe. There were plant gatherers, huntsmen and butterfly catchers and, out in front, a herald and two messengers with the baron's initials emblazoned on their breastplates. Over all there was a *sirdar* of forbidding countenance with a shield on his back and a sheathed sabre forever in his hand.

And, of course, there was Vigne. The Austrian was too correct and the Englishman too easy-going for an open quarrel. But Vigne was always 'giving way to his servants' and 'always tardy'. When it rained he stayed in bed while the baron fussed and fretted in the mud. His high spirits were incompatible with the baron's dignity and his down-to-earth common sense was poor company for an aspiring soul oppressed by world weariness. By the time they reached Lahore the baron, 'heartily tired of solitude', longed for the company of a European as if his companion no longer existed.

Ranjit Singh, though now an old and sick man who spent much of his time closeted in his zenana, realised that von Hugel was of higher rank than most of his previous visitors and determined to impress him. The court was now at its sumptuous zenith, and there were treasures galore to be paraded. The Koh-i-Noor diamond, filched from the Afghans and described by Vigne as the size of a walnut and by von Hugel as like a hen's egg, prompted the baron to ponderous reflections on the value of worldly goods. Leili, the celebrated steed which had cost the Maharaja the despatch of an army, was also pointed out, though Vigne doubted whether they were seeing the real animal. Every night the Maharaja sent over delicacies from his kitchen and dancing girls from his harem. The baron was not much interested in women nor in horses but on military matters he could hold his own with anyone. Special manœuvres were held in his honour and one gorgeous parade followed another.

Everything that belonged to the Maharaja was festooned with jewellery. The nose rings of his Kashmiri *nach* girls fell to below their narrow waists in a cascade of diamonds, pearls and emeralds. The saddlery of his horses was aglow with precious stones; there were cruppers of emeralds, reins of gold thread and saddle cloths dripping with pearls and coral. The howdahs of the elephants were of solid gold or silver and their trappings set with diamonds and turquoises. The royal tents were made of Kashmir shawls and on ceremonial occasions the troops were dressed in yellow satin with gold scarves or in cloth of gold patterned with scarlet and purple. Every sword and matchlock, shield and spear was set with gems.

It was indeed one of the world's greatest collections of treasure. Some of it was old, the accumulation of Indian, Afghan and Persian conquerors. But the visitor was struck by the fact that nothing was dull or faded. The colours were bright and the gems newly set and polished. It was not the legacy of some defunct glory but had been fought for and won by the men who displayed it. The guard of honour might look like 'a row of gaudy and gigantic tulips' but over the yellow satin flowed long grey beards and beneath the gold breastplates were scars still livid.

Before this mouth-watering display von Hugel hesitated. All along his thoughts had been focused on the Bombay sailing and his arrival back in Vienna. Now the Maharaja was offering him a share of all this if he would accept employment as one of his generals. Visions of untold glory rose before him. He saw himself leading his own army into Central Asia; 'by one man's efforts civilisation might be mightily advanced'. Then as so often, he remembered his old mother back home in Austria. Whatever the cost to civilisation it was unthinkable to desert her. He declined the offer and soon after left for Ludhiana and Europe.

Vigne in all this was just a passive observer. Military matters bored him but, so as not to prejudice his chances of returning to Kashmir, he played the role of an appreciative and obliging Boswell.

His only error was when, to him too, the Maharaja offered employment. What did he think of the governorship of Kashmir? Remembering Jacquemont and anticipating the offer, he smiled. When it came he laughed outright. Unlike the baron, he realised that this was just one of Ranjit's usual compliments. No one became a general or governor overnight. Men like Allard and Avitabile had sought and fought for their commands.

In the course of three visits to Lahore Vigne became increasingly disrespectful of the Sikh court. He began to notice the sordid as well as the sumptuous. The Maharaja never washed and his idea of a practical joke was to urinate—for reasons of delicacy Vigne tells the story in Latin—from an elephant on to the turbanned heads of his subjects. He was not given to senseless cruelty but as fast as he might exalt one man he would crush another for the pettiest of reasons. As for his courtiers they were all 'blackguards . . . to whom a disregard of principle, subtle intrigue and calm hypocrisy were alike familiar and diurnal'.

In 1837, after the rough handling in Leh, he returned to Lahore burning with indignation. It was clear to him that the man behind his troubles was not the Maharaja but Gulab Singh, the powerful courtier and Raja of Jammu. This was the man who had been persecuting Jacquemont's bandit chief and who was now well advanced with his designs on the whole mountain region. Vigne was among the first to realise that he was already virtually independent of Lahore. He and his brothers held Jammu, most of the Pir Panjal and Ladakh. The addition of Baltistan would mean the complete encirclement of Kashmir. The Raja was only biding his time until the death of Ranjit Singh, when he would grab the valley and declare his independence.

Vigne had some respect for the old Maharaja but none at all for Gulab Singh. He was a cruel and scheming tyrant who skinned his prisoners alive and was hated by his subjects from Tibet to the Punjab. More specifically, it was on his instructions that Vigne had been manhandled in Leh and the *firman* from the Maharaja thus flouted. Ranjit promised redress but the power of Gulab Singh was not curbed. To Vigne it seemed that it never would be until Ranjit was dead, the Punjab and Kashmir annexed by the British and all the hill states, including Ladakh, handed back to their traditional rulers.

By now, the late 1830s, conjecture about the future of the Sikh

empire after Ranjit's death was general. It was taken for granted
that it would disintegrate. There was no obvious and able successor
and too many immensely powerful aspirants. It also seemed inevitable that the British would be drawn across the Sutlej. As a contemporary writer put it, 'The East India Company has swallowed
too many camels to strain at this gnat.' In 1839 the Maharaja finally
bowed out; his body was drawn to the pyre on a ship made of gold
with cloth of gold sails to waft him off to Paradise. Ten years and
two hard fought wars later—the gnat had quite a sting—Lahore was
in British hands. But the mountains from Jammu to Skardu and
from Tibet to Muzafarabad were united under the rule of the first
Maharaja of Kashmir. Vigne at this time was in London planning an
expedition to Central and South America. He died at Woodford,
only a few miles from where he was born, in 1863, never having
returned to India or Kashmir. It was hardly surprising. The first
Maharaja of Kashmir was none other than his arch enemy, Gulab
Singh.

PART TWO

The Pamirs and the Hindu Kush
1826–1841

He took the one to the mountains,
He ran through the vale of Cashmere,
He ran through the rhododendrons,
Till he came to the land of Pamir.
And there in a precipice valley,
A girl of his age he met,
Took him home to her bower
Or he might be running yet.
ROBERT FROST. *The Bearer of Evil Tidings*

6. Running Gun

With the publication in the early 1840s of the travels of Moorcroft and of G. T. Vigne, exploration in the Western Himalayas came to be concentrated on opening a north–south route from India to Turkestan via Ladakh. When the mountains were seen as a barrier, this looked like their weakest point. Whether it could be forced from India for trading purposes, or from the outside by an invading force, became a matter of deep concern in British Indian policy.

But, seen as a knot of mountains constituting a purely geographical challenge, they could as well be tackled from other points of the compass. Tibet in the east and Chinese Turkestan in the north were out of the question, but the approach from Afghanistan in the west was still technically feasible. This was the route taken by both Marco Polo and Benedict de Goes on their way to the Pamirs, and one of the busiest trade routes from Bukhara to China followed the same axis. Moorcroft's and Wolff's experiences in Afghanistan were not exactly encouraging but here there was no Ranjit Singh holding the keys to the mountains and it was too far from the British frontier for the Hon. Company to exercise a restraining influence.

Thus, while Jacquemont, Wolff and Vigne were trying to get beyond Kashmir, two attempts were made to penetrate the mountains from this direction. They were in no way connected, but, together, they shed new light on the whole mountain complex and particularly its western ramparts, the Pamirs and the Hindu Kush. The first of these journeys, mysterious, improbable and little known, was generally reckoned the less important. Geographers for the most part could make nothing of it; politicians and strategists seldom even knew of it. Yet the wanderings of Alexander Gardiner, given their date and circumstances, are perhaps the most remarkable in the whole field of nineteenth century travel in Asia. Single-handed, without official support and without any geographical training, this man had, by 1831, already explored the Western Himalayas.

Ten years before Vigne gave up his attempt to reach Gilgit, Gardiner had been there. Twenty years before Thomson tackled the Karakoram pass, Gardiner had crossed it. Forty years before

Shaw and Hayward reached Yarkand, Gardiner had passed through the city. And fifty years before an Englishman reached Kafiristan, Gardiner had returned there a second time. He had crossed every one of the six great mountain systems before the maps even acknowledged their individual existence and he had seen more of the deserts of Turkestan than any non-Asiatic contemporary.

But to win acclaim as an explorer, to enjoy the publisher's royalties, the medals and the honours, it is not enough just to have travelled. The successful explorer must interrupt his movements to take measurements and observations. He must carefully identify physical features and place names and, at the end, he must write a convincing, coherent and consistent report. On all these counts Gardiner failed. His travels are not lacking in detail. There are names galore and there is a wealth of colour and excitement; perhaps too much, for though the appearance of authenticity is irresistible, the substance is sadly lacking. The directions, distances and dates don't seem to tally, the place names and weird peoples are either non-existent or unidentifiable and his occasional companions are almost invariably men unknown to history. The one event, the one encounter, the one concrete fact or observation which would clinch the veracity of Gardiner's story is missing.

This aura of uncertainty surrounds not only his travels but his whole life. Who Gardiner was, where he came from, how he reached Central Asia, what part he played on the political scene and what schemes he was hatching at the very end of his long and eventful life, are all questions that he himself answered with the same tantalising disregard for supporting evidence. Just as there were those who could not accept the story of his travels, there were some who doubted whether he even existed. But this at least can be proved. There is the evidence of those who met him and, above all and most unexpectedly, there is a photograph of him.

'A kenspeckle figure' is a description much beloved by Scottish obituary writers. It implies that the deceased was well known to the point of being conspicuous, and that somehow his home town will never be quite the same without him. In Srinagar in the 1860s and 1870s Alexander Gardiner was a kenspeckle figure. He was six feet tall, thin and old, but he stood very straight, stared very hard and his moustaches and whiskers curled defiantly upward. No one knew quite where, or in what style, he lived. It was somewhere deep in the bowels of the city where no European—and by then there were many in Srinagar—ever penetrated. When, to satisfy the curiosity of a British visitor, he emerged from his lair, it was obvious that he had 'gone native' many years before. His uniform consisted of a turban crowned with heron plumes which, in the armies of Punjab and Kashmir, denoted rank; Gardiner called himself a Colonel. The jacket and trousers were of European style but markedly native cut. And the whole ensemble, from toe to turban, was of tartan plaid. It was acquired, according to the cognoscenti, from the Quartermaster's Stores of the 79th Highland Infantry.

Stretched in one of those long cane chairs which were an essential piece of furniture in a British home in India, Gardiner would start on his tales. The only acceptable interruption would come when he paused for a drink. His host, in embarrassment, would then fiddle with his pipe or turn to toss another log on the fire. The old Colonel had difficulty in swallowing. His now frail body bore the marks of fourteen ghastly wounds, of which the most inconvenient was in his throat. He could only manage liquids and, to get even these down he had to clamp, with one hand, a pair of steel pincers round his gullet while, tipping his glass with the other, he gulped painfully.

At first it was not easy to follow his story. His English was rusty with disuse and his expressions distinctly archaic. He was also tooth-less and had what sounded like an Irish accent. But once he got going, the most critical listener would be lulled into rapt attention. The photograph 'gives but a dim idea', according to one of them, 'of the vivacity of expression, the play of feature, the humour of the mouth and the energy of character portrayed by the whole aspect of the man as he described the arduous and terrible incidents of a long life of romance and vicissitudes'. However shaky his English, he knew how to tell a good story. With those staring eyes ablaze, moustaches flailing the air and surprisingly elegant hands holding the thread of his tale, he would unravel the extraordinary jumble of far-away places and unexpected encounters that go to make up his life story.

It begins in the United States in the year 1785. To a Scottish doctor and his half-English, half-Spanish wife, a third son was born on the shores of Lake Superior not far from the source of the Mississippi. He was christened Alexander Haughton Campbell Gardiner. Five years later the doctor entered the Mexican service and the family moved to near the mouth of the Colorado river and a town called St. Xavier. There the children were educated in a Jesuit school, though the young Alexander preferred to devote his time to a book of pioneer travels amongst the North American Indians. Already 'the notion of being a traveller and adventurer and of somehow and somewhere carving out a career for myself was the maggot of my brain'.

Recalling his youth all of sixty years later, Gardiner was understandably hazy. But in locations like 'the shores of Lake Superior', 'the source of the Mississippi' and 'the mouth of the Colorado river', he gives an early indication of an infuriating inexactitude. Moreover, no map has ever marked a place called St. Xavier.

About 1809 he went to Ireland. He returned to America in 1812 but found his father dead and immediately headed back to Europe. He was ever proud of his American citizenship, but never again crossed the Atlantic. In Spain he realised the estate of his mother and then, via Cairo and the Black Sea, reached Astrakhan on the Caspian. Here his eldest brother had settled down as a mining engineer in the employ of the Russian government. The young Gardiner hoped to follow in his footsteps and spent three years immersed in mineralogy. In 1817 the brother died, the Russian authorities sequestered most of his property and Alexander was refused employment. Disgusted, he packed his bags and set sail across the Caspian. He was thirty-three.

For the next thirteen years Gardiner lived in the saddle. He wandered from Ashkhabad to Herat, then amongst the Hazaras and Turkomans to Khiva, north to the Aral Sea and back across the Caspian to Astrakhan. That was just the preliminary. In 1823 he set off again, across the steppes from the Caspian to the Aral and then up the Syr river, the Jaxartes of the ancients, to Ura Tyube near Samarkand, a distance of over 1,000 miles. By now he had decided that he 'could not rest in civilised surroundings'. He was happy only amongst wild races and unknown lands. He passed by the name of Arb Shah, wore Uzbek costume and carried a Koran into which he stuffed the scanty notes of his travels.

He had also acquired a band of faithful followers. There was no law in Central Asia. Outside the cities it was every man for himself. The distinction between the predators—robbers, slave dealers and nomads—and the prey—merchants, pilgrims and herdsmen—was a fine one indeed. But, as far as Ura Tyube, Arb Shah and his friends had moved with the circumspection of bona fide travellers. Now, after a complex three-sided engagement in which they lost and then stole back their horses and property, they became fugitives. Previously they had avoided cities for fear of Gardiner's nationality being detected. Now they had no choice in the matter. They were outlaws riding by night, hiding by day and living on what they could steal.

At Hazrat Imam they crossed the Oxus into Afghanistan and headed for Kabul. Ever since leaving Astrakhan, Gardiner had had the idea of enlisting in the forces of an Asiatic ruler. The Shah of Persia, the Khan of Khokand and Ranjit Singh had all figured in his plans. In Afghanistan his best chance lay with Dost Mohammed, now establishing himself in Kabul, and he hoped that, in return for his services, he and his followers would be granted an amnesty. There is a distinguished tradition, particularly in India, of European military adventurers, men who, departing or deserting their national colours, served in the armies of native princes. Jacquemont's friend Allard and Wolff's 'kind Italian' Avitabile, are notable examples. Much later Gardiner did join them in the Sikh service, but to compare him and his desperate little band, as they came fleeing out of Turkestan, with Napoleonic officers at the head of well-trained armies, is misleading. Better by far to remember his childhood in that mysterious town somewhere on what is now the Arizona/Mexico border. Gardiner was just a desperado, a hired gun on the run. His moral standards, which later caused such a scandal, were those of Boot Hill, not Sandhurst. If, instead of a tartan turban, the old Colonel had worn a ten-gallon hat and if, instead of Srinagar, his resting place had been St. Louis, his whole career would have been more readily understood.

To avoid detection they gunned down three of Murad Beg's men near Kunduz, and would have done the same again when stopped in the Kohistan just north of Kabul. Only this time there were fifty picked horsemen who came on them 'like a desert storm'. Gardiner drew up his men and prepared to parley. His captor was Habib Ullah Khan, the dispossessed heir to the throne of Kabul, who was now waging the nineteenth century equivalent of a guerilla war against his uncle, Dost Mohammed. History has condemned Habib Ullah as a debauchee, a drunkard and a coward. Gardiner, on the other hand, painted him in heroic shades as a born leader of dauntless courage, high-minded principles and magnanimous intentions. This was important, for it struck those who wished to discredit his story as highly significant that, in his one supposed encounter with a figure known to history, he could be so demonstrably wrong.

In fact he may not have been. Posterity's verdict on Habib Ullah was that given out by his arch enemies, Dost Mohammed and his

brothers, and substantiated by the Khan's depravities after his final defeat. The only foreigner who met him in the days of his youth, apart from Gardiner, was the archaeologist, Charles Masson.*
Writing in the light of what later befell Habib Ullah, he too calls him 'rash, headstrong, profuse and dissipated'; but in his youth he found him a splendid figure. He dressed in gold and scarlet and was the finest looking man in the whole of the Afghan court. His vices were 'rather those of habit than of the heart and to atone for them he possessed indomitable personal bravery and lavish generosity'.

Such qualities were no match for the scheming duplicity of his uncle, but one can understand why they appealed to a man like Gardiner. Instead of a choice between death and slavery, the generous prince offered him command of 180 horsemen in his struggle against Dost Mohammed. Gardiner accepted and, for the next two and a half years, led a life of 'active warfare and continual forays . . . for the good cause of right against wrong'. From the mountain fringes between Jalalabad and Bamian they swooped into the settled lands of the Kabul river basin or waylaid caravans crossing the Hindu Kush. According to Masson they were 'eight hundred very dissolute but resolute cavalry living at free quarters upon the country'.

It was a far from settled existence but, amid such excitements, Gardiner enjoyed his first tragic taste of domestic happiness. In an attack on a caravan of distinguished pilgrims he had glimpsed the face of a beautiful young girl. As his share of the booty he asked for and, though she was of royal blood, was given this maiden. They set up home in a fort near Parwan and 'there I was happy for about two years in the course of which time my wife made me the father of a noble boy'. This was the beginning of the most difficult part of his narrative. The tone becomes subdued. Forty years had elapsed but still the tears flowed as he recalled it. 'I must hurry over this part of my story.'

By 1826 Habib Ullah's cause was lost. He was pinned down near Parwan and Gardiner was ordered back from his forays. When he arrived, the Khan's forces had already been defeated; Habib Ullah

* The American Charles Masson was originally the Englishman James Lewis of the Bombay army. He deserted in the 1820s and wandered off into Afghanistan where he spent most of the next fifteen years. Travelling in the style of Dr. Henderson, unarmed, on foot and penniless, he gained a deeper understanding of Afghanistan than any of his contemporaries. He first won attention in India with his archaeological reports on the Buddhist topes around Kabul, and in 1835 he was appointed British informant in that city

himself, though still fighting, was severely wounded. Gardiner hacked a path through to his side only to learn, when the enemy retired, that his own fort had already been stormed. What he found, when after a frantic ride he entered his home, is best left to the old Colonel himself.

The silence was oppressive when I rode through the gateway of the fort, and my men instinctively fell back, when an old *mullah* (who had remained faithful to our party) came out to meet me, with his left hand and arm bound up. His fingers had been cut off and his arm nearly severed at the wrist by savage blows from a scimitar while striving to protect my little child. Faint from his wounds and from the miserable recollection of the scene from which he had escaped, the sole survivor, the aged *mullah*, at first stood gazing at me in a sort of wild abstraction, and then recounted the tale of the massacre of all that I loved.

The garrison had long and gallantly held their own, though attacked on all sides by an immensely superior force. They had seen Habib Ullah approaching, fighting gallantly, and had for a moment thought themselves saved, but he had been driven back and passed from their sight. The *castello* had been stormed and all in it put to the sword, with the sole exception of the old priest.

After this brief story the *mullah* silently beckoned me to dismount and to follow him into the inner room. There lay four mangled corpses—my wife, my boy and two little eunuch youths. I had left them all thoughtless and happy but five days before. The bodies had been decently covered up by the faithful *mullah*, but the right hand of the hapless young mother could be seen, and clenched in it the reeking *katar* with which she had stabbed herself to the heart after handing over the child to the priest for protection. Her room had been broken open, and mortally self wounded as she was, the assassins nearly severed her head from her body with their long Afghan knives or sabres. The *mullah* had tried to escape with the child, but had been cut across the hand and arm as aforesaid, and the boy seized and barbarously murdered. There he lay by the side of his mother.

I sank on my knees and involutarily offered up a prayer for vengeance to the most high God. Seeing my attitude the *mullah*, in a low solemn tone, breathed the Mohammedan prayers proper to the presence of the dead, in which my *sowars*, who had silently followed with bent heads, fervently joined. Tear after tear trickled down the pallid and withered cheeks of the priest as he concluded. Rising, I forced myself and him away from the room, gave him all the money I had for the interment of the dead, and with fevered brain rode away for ever from my once happy mountain home.

It was the end of his Afghan adventure and the beginning of his journey into the Western Himalayas.

The inventory of essential equipment and provisions that travellers feel constrained to present to their readers soon become dull. Those 'double fly tents', 'artificial horizons', 'tinned' cooking pots and all the endless arguments about boots versus the native grass shoes invariably presage a second-rate narrative with little pioneering content. The great explorers tend to be more reticent and less dogmatic about their preparations. Gardiner is no exception. His party of desperate men who now struck off up into the Hindu Kush, assessed the total value of their property at just over half a rupee.

They were not, of course, consciously embarking on an expedition; they were running for their lives. Gardiner himself was wounded in the neck and the leg. Washed in salt and water and dressed with powdered charcoal and clay, it was not surprising that these wounds never really healed. His outfit consisted of a high black Uzbek hat, black sheepskin coat, hair rope girdle and Turki boots. They had guns and horses but for food depended on what they could steal. Until the first well-provided caravan passed within their grasp, they lived off 'snow mushrooms' and a half rotten, partially cooked 'hyena-like animal'.

The plan at this stage was a simple one, to put as much of Afghanistan as possible between themselves and Dost Mohammed. They climbed north into the mountains, crossing by the Khawak pass and hoping once again to escape over the Oxus at Hazrat Imam. This great river, which divides Afghanistan from Turkestan proper was regarded by Gardiner as a second Rio Grande; safety always lay on the other side. The problem was that between the Hindu Kush and the river lay a hundred miles of open ground, the domain of Murad Beg. It was only two years since Moorcroft had been waylaid there and the Uzbek chief was known to be keeping a close lookout for stragglers from Habib Ullah's forces.

The first attempt to descend from the mountains was a disaster. A group of horsemen quietly crossing the plains below them suddenly changed direction and came on at full gallop. They were fifty strong. Gardiner's force, augmented by more stragglers, numbered thirteen. The only chance of escape lay over a pass which he calls Darra Suleiman and in the best Western tradition they tore off towards it. Fast and furious was the pursuit, but Gardiner's men were stretching the gap. Then, half a mile from the pass and safety, another smaller group appeared from among the rocks and barred their path.

The fray now became general, as the main body charged us, trying to save their comrades. This fortunately prevented their using their matchlocks, and we had reached the mouth of the pass, which we held with desperation. Their overwhelming numbers, however, soon broke our ranks, and they unfortunately got mixed up with us; there was no room for orderly fighting and it was a mere cut and thrust affair.

Soon we had only seven men left out of thirteen, and we slowly retreated up the pass, keeping them off as well as we could. In the pass we lost two more men and were now reduced to five, each of us severely wounded. I myself received two wounds, one a bad one in the groin from an Afghan knife and the other a stab from a dirk in the chest.

It was now quite dark and the rain was coming down still heavier than before. However our enemies followed us no further— no doubt the plundering of the dead being their chief inducement to return. We made our way through the pass as quickly as we could in the midst of heavy rain, hail and lightning, while the roll of thunder seemed to make the very rocks around us and the ground beneath us to vibrate most sensibly. What with my two former wounds still raw, and my two fresh ones (one of which was bleeding freely) I was soon so weak as to nearly faint in my saddle.

There was no further pursuit and, after a few days' recuperation in the mountains, they headed north-east into Badakshan. They were still making for the Oxus but now planned to cross it much further east where it emerges from the Pamirs. Above Jerm they descended into the valley of the Kokcha tributary, and this time there was no attack. The country had recently been depopulated by Murad Beg. After a change of travelling companions and more rest, Gardiner finally crossed the Oxus opposite the Shakhdara and entered the valley of Shignan. He was now on the threshold of the Pamirs.

The year was still 1826. In the surviving accounts of his story the next date is spring 1830, by which time he has crossed the Himalayas from north to south and is incarcerated in a dungeon five hundred miles away in Kandahar. As far as the Pamirs his wanderings have been full of detail though not all of it credible; there has been an explanation for each change of direction and, with a good map, the gist of his itinerary can be followed. Now suddenly, at what is geographically the crucial stage, his account becomes vague and disjointed. Did he actually cross the Pamirs or skirt them by way of the foothills of the Alai, two hundred miles to the north? It is impossible to say. He was aiming for Yarkand but until he actually got there none of the places he mentions can be positively identified.

The Pamirs are a polar wilderness combining the bleakness of Tibet with the ruggedness of the Karakorams. Descriptions of them are often unsatisfactory for the simple reason that there is little to describe. In summer Kirghiz nomads go in search of the grazing, but no one actually lives there. The sum total of features recorded by the first dozen visitors is two or three crumbling tombs and a dreary muddle of lakes and rivers. It is not, therefore, altogether surprising that Gardiner was unable to substantiate his claim to have been there. On the other hand one would expect some mention of the two most obvious peculiarities of the region, the cold and the wind. Lord Dunmore, who crossed the Roof of the World in 1892, reckoned it the coldest place on earth. He should have known: he had already had a taste of Spitzbergen, the Hudson Bay and Arctic Russia. At night his thermometer registered 45° of frost, and this was inside his tent. Coupled with the debilitating effect of an altitude never less than 12,000 feet, it made life a precarious business. But the wind added a whole new dimension of discomfort. Capricious in direction, unpredictable in strength and unimpeded by anything approaching shelter, it blew day and night, summer and winter, on the mountains and in the valleys, with a vicious numbing intent. It was not a lusty gale to be ridden out, nor a steady blast that one learned to live with; but more like a nagging and insidious cancer. An unprotected face it would cut to shreds in minutes; over a period of days it ate deep into the spirit of a man. Sapping his energy and numbing his senses, it claimed its victims almost unnoticed.

Now Gardiner omits all mention of this. When sixty years later the Russians established their first post on the Pamirs they found the conditions too harsh to support a normal existence. Their poultry all died and the only crop that succeeded was a woody radish. Gardiner, on the other hand talks of Shignan as a veritable orchard, a description that could only apply to the westernmost fringe of the region where it borders the Oxus. He says nothing about the altitude, the cold or the snow-clad peaks. A journey of forty miles which took them all of seven days could just as well apply to the rugged country along the banks of the Upper Oxus.

Instead, he had stories of a grizzly fight with a pack of wolves, a visit to the ruby mines on the Oxus and a lot of fraternisation with the Kirghiz herdsmen of the area. There is also some questionable information on aboriginal tribes in the Alai and a ruined city near Yarkand with 'chasms and caverns . . . encrusted with corrosive salts and pervaded with mephitic vapours'. Sometimes he can be positively faulted. No yak ever survived in the deserts of Turkestan. His description of them is convincing enough and almost certainly he would have seen them in Ladakh or on the Pamirs; it looks as if his memory has simply transposed them. Moreover no one would question the old-timer's right to embroider his stories. What was significant was that something like his description of a Kirghiz wedding could be so unexpectedly accurate.

Yarkand, the first recognisable placename, was reached on September 24th. The year is anyone's guess—it could have been 1827, '28 or '29. Here the treatment again changes. A traceable itinerary re-emerges but the colourful detail is gone. From Gardiner one expects classic descriptions and harrowing predicaments amongst the mighty peaks and glaciers. But there is nothing. Just:

21st. Along banks of river to Bolong Belook.
22nd. Ditto to where Doorg meets it on other side.
23rd. Ditto to Fitkar.
24th. Ditto to Lohoo.
28th. Arrived at Leh.
30th. Leave Ladakh for Cashmere.

And so on. From his description 'the worst trade route in the whole wide world' could just as well be a pleasant country by-way and the Karakorams no more formidable than the Hog's Back.

Gardiner's omissions are really more serious than his doubtful assertions. The austere itinerary above is taken from his 'journals'. These never seem to have constituted more than random jottings and, without the accounts of those who heard the story from the

Colonel himself, they are practically meaningless. Gardiner explained that some of the journals were being studied by Sir Alexander Burnes and perished with him in Kabul in 1841. This loss, he claimed, accounted for any inaccuracies or gaps in his story. But in this particular case the journal was not lost and, even if it had been, it is hard to believe that he could recall nothing of his crossing of the mightiest mountains in the world. Besides which, Gardiner's imagination was not usually behindhand in making good any deficiencies of detail. Was it, one wonders, that this section of his travels was one which by the late 1860s could too easily be verified? Or, more charitably, does one assume that he regarded as superfluous any elaboration on his part of a route now widely publicised?

Seven weeks after leaving Yarkand he rode into Srinagar. It was good going but, for a party travelling light during the favourable autumn months, not unreasonable. Since leaving Afghanistan Gardiner had advanced no explanation of where he was heading or why. He had in fact prescribed a massive arc, north to Shignan and perhaps the Alai, east to Yarkand, south to Leh and now west to Srinagar. Here reports that Habib Ullah was again in the field determined him to complete the full circle by returning to Afghanistan. The easiest route would have been that followed by Moorcroft via the Punjab, but for some reason Gardiner and his men stuck to the mountains. They passed through Chilas, Gilgit and Chitral, and finally entered Kafiristan. These are all names which figure prominently in a later phase of exploration in the Western Himalayas. In the 1820s neither their existence nor position was certain and, even when the old traveller was spinning his yarns fifty years later, there were still few, if any, first-hand accounts of them by Europeans.

They are also the only names in his itinerary that can be identified. The rest, and in the published extracts of his journals there are many —villages, forts, 'Cyclopean' ruins—are a mystery. A footnote by the editor of the journals explains that, where the handwriting permitted, Gardiner's spelling had been scrupulously followed. But allowing for bad writing, imperfectly heard names and an erratic system of transliterating them, the reader may take the ultimate liberties in juggling with consonants and still only a handful more can be positively identified.

On the other hand, on this final stage of his journey through the mountains, there is at least some attempt to describe the terrain. Gardiner had no idea of the structure of the mountain complex and anyone who has gazed, mapless, on the chaotic contours north and west of the Indus will sympathise with him. A high pass may be taking one over one of the great Asian watersheds and therefore the spine of one of the main ranges. Or it may be just an insignificant spur. Without the time to explore each valley and follow every stream there is no way of telling. Gardiner was still moving at speed. He clearly relied a good deal on native reports for his general observations but there is sufficient mention of snow-fed torrents, steep ascents, precipitous ledges, mountains covered with perpetual snow and avalanches to make it all sound a good deal more convincing than his oblivion about the Karakorams.

The visit to Kafiristan was, in retrospect, the crowning achievement of his travels. Even today very little is known about this mountain fastness between Chitral and Afghanistan. Gardiner was the first man to claim to have visited it, though he reports that two Europeans had been there some sixty years before. This casual observation lends much credibility to his story, since exactly the same legend was uncovered by a British mission to the area in the 1880s. It was upon their report that Kipling based his story of 'The Man Who Would Be King'.

Gardiner regarded this as his second visit. The first was when he went some way east up the Hindu Kush while fleeing from Dost Mohammed. He says that he was then offered the command of a Kafir tribe and, however this may be, he does seem to have struck up a valuable friendship with this normally hostile people. Mention of them occurs so often in his stories that one is tempted to infer a relationship similar to that between Moorcroft and the Ladakhis or Vigne and the Baltis. On the second visit he adopted the goatskin wrapping of the Kafirs and travelled alone and on foot. He penetrated much further and was probably there much longer than on the first occasion. Even without the relevant journal he was later able to give a more detailed account of the place than anywhere else on his route through the mountains.

Of all the Himalayan peoples the Kafirs are the most distinctive, the most curious and the most primitive. Travellers who met them as slaves in Afghanistan were first intrigued by their appearance. A Kafir girl noticed by Sir Henry Rawlinson in Kabul was the most beautiful oriental he had ever seen. She was the only lady he had ever met who, by loosening her tresses, 'could cover herself from head to foot as with a screen'. Gardiner, who never fails to comment on the female section of the population, describes the women as having hair 'varying from the deepest auburn to the brightest golden tints, blue eyes, lithe figures, fine white teeth, cherry lips and the loveliest peach blossom on their cheeks'. Not only did they look a bit European but they had certain European customs. Marooned in a sea of peoples who habitually squat on the ground, the Kafirs sit on chairs. They make and drink wine and their dead they put into coffins.

This was not, however, enough to convince the first European visitors, Gardiner included, that they should be acknowledged as brethren. Appearances could be deceptive. The Kafirs were fascinating but they were nearer to savages than any other people in Asia. They were appallingly dirty and immoral, their language was unpronounceable and their religion the most primitive animism. Burnes records that, when a Kafir raiding party returned from a foray against their Mohammedan neighbours, it was usual for the warriors to bring back the severed heads of their enemies. Any luckless brave who came back empty-handed was debarred from the homecoming celebrations. These commenced with a ritual hunt for walnuts, the nuts having been previously hidden by the young ladies of the tribe in their bosoms. The story sounds suspiciously like one of Gardiner's. Burnes was certainly supposed to have had the journals of Gardiner's stay in Kafiristan, and more's the pity they perished with him.

Emerging into more civilised surroundings at Jalalabad, Gardiner finally completed his tour of the Western Himalayas. So far as is known he never again visited the lands beyond Kashmir, and the rest of his story can be briefly dealt with. Habib Ullah had not reappeared. Gardiner, trying to reach Persia, was imprisoned at Kandahar, escaped and eventually applied again to enter Dost Mohammed's service in Kabul. Not surprisingly he was refused but at Peshawar he found employment with one of Dost Mohammed's brothers and from there, in 1831, gravitated to Ranjit Singh's army as an artillery officer. He was one of the few foreigners who remained in the Sikh service after Ranjit's death and he played a prominent role in the extraordinary events that followed. These, as Gardiner's biographer has it, amounted to 'a rapid succession of crimes and tragedies such as have rarely been paralleled in history save in the darkest period of the downfall of Rome or the early days of the French revolution'. Gardiner himself listed nineteen participants in the power struggle. Of these, in the five years 1839–44, sixteen were murdered or died under suspicious circumstances. Events, for once, outstripped even the Colonel's lurid imagination. The facts were more macabre than fiction and his account of them has been largely substantiated. From an historical point of view it is the most valuable part of his whole tale. Of the three surviving contenders only one succeeded in retaining any part of Ranjit's empire. This, of course, was Gulab Singh, created by the British, Maharaja of Kashmir. Gardiner disliked him as heartily as did Vigne. His character was 'one of the most repulsive it is possible to imagine', and he accused him of barbarities and atrocities perpetrated in cold blood 'for the sole purpose of investing his name with terror'. Yet, only months after this damning attack, he enlisted in Gulab Singh's service and in Srinagar, as commandant of the new Maharaja's artillery, spent the rest of his active career and then a long and honourable retirement.

7. A Most Distinguished Old Man

Such, then, was Gardiner's story. He died in his bed—after such a life, no mean achievement—in 1877 aged, by his own reckoning, ninety-one. It was the end of one saga but only the beginning of another. Already a few geographers were struggling with his journals and wondering whether they were on to the greatest journey since Marco Polo's or one of travel's most elaborate fabrications. Others, in the interests of history, were trying to piece together the accounts of his life. They at first had fewer reservations, though eventually it was an historian who published the most devastating rebuttal of all.

The first and only printed extracts of Gardiner's journal appeared in India in 1853, with an introduction by Edgeworth, a Bengal civil servant. Lord Strangford of the Royal Asiatic Society drew them to the attention of the Royal Geographical Society, and it was thus that the two great umpires of Asian travel, Sir Henry Rawlinson and Sir Henry Yule, first heard of Gardiner. They only had the journal extracts to go on and these are, as noticed, semi-literate jottings, to this day unintelligible.

'Geography,' declared Yule, 'like Divinity has its Apocrypha . . . I am sorry to include under this head the diary of Colonel Gardiner.' Every attempt to construct from it a consecutive itinerary had failed. The recognisable names, Kunduz, Badakshan, Yarkand, Gilgit etc. were too few. 'But amid the phantasmagoria of antres vast and deserts idle, of scenery weird and uncouth nomenclature, which flashes past us in the diary till our heads go round, we alight upon these familiar names as if from the clouds; they link to nothing before or behind; and the traveller's tracks remind us of that uncanny creature which is said to haunt the eternal snows of the Sikkim Himalaya and whose footsteps are found only at intervals of forty or fifty yards.' He stopped just short of denying that Gardiner had ever visited those places, but on the evidence he thought it unlikely.*

* Gardiner's travels were approached with excessive caution because their currency in the 1860s and '70s happened to coincide with a case of forged maps and fabricated itineraries also relating to the Pamir region. Known as

Rawlinson was more cautious. He regarded the journals as so mutilated and exaggerated 'that the narrative reads more like a romance than a journal of actual adventure'; their author was 'untrustworthy'. On the other hand he believed that 'Gardiner did certainly visit Badakshan' and 'actually traversed the Gilgit valley from the Indus to the Snowy Mountains and finally crossed into Chitral being in fact the only Englishman [?] up to the present time ever to have performed the journey throughout'. Whereas Yule in 1872 imagined Gardiner dead, Rawlinson, writing in 1866 and again in 1874, knew that he was still alive. He knew too that there were others working on the Colonel's verbatim accounts and he fancied that Gardiner, now an older and soberer man, would yet be vindicated.

In this he was right. In Kashmir the old soldier in the tartan uniform was winning support from everyone who met him. These included Frederick Cooper, a member of the Indian Civil Service who was deputed to Kashmir in 1864. He undertook with Gardiner's assistance to write a definitive biography. The work, corrected and therefore endorsed by the Colonel himself, had progressed only as far as the Pamirs when Cooper died in 1871. This helps explain why the account of the rest of his travels through the mountains is so unsatisfactory. Gardiner was never able to contribute directly to it for, with Cooper's death, the draft and notes promptly disappeared and were not rediscovered till after Gardiner's death.

Sir Henry Durand, Lieutenant-Governor of the Punjab, visited Srinagar in 1870 to enquire into allegations brought against the Maharaja by another Himalayan explorer, George Hayward. While there he too met Gardiner and declared him 'one of the finest

---

the Klaproth forgeries from the German geographer who was allegedly responsible for them, these had the savants of the geographical societies of London, St. Petersberg and Paris locked in heated debate. The subject is a highly complicated one and is treated ad nauseam in the Societies' journals from 1865 to 1875. The forgeries had found their way into most maps of the region and had perpetuated an error originally made by the Jesuit cartographers of the eighteenth century. The fathers had managed to twist the section of their map which dealt with the Pamirs through 90 degrees so that, when mounted in an overall map of China and its confines, the Oxus ran north–south instead of east–west. Poor Gardiner, when he tried to reconstruct his travels with the help of a map, was using one that was not just inaccurate but hopelessly falsified. In view of this it is hardly surprising that the account of his journey through or round the Pamirs is so vague. He must have been as confused about his itinerary as anyone.

specimens ever known of the soldier of fortune'. He drafted a short article for the Indian press, *Life of a Soldier of the Olden Times*, but, like Cooper, died before it could be published. (He was actually knocked off the back of an elephant by a low archway whilst on a ceremonial visit to the state of Tonk.) Fortunately his son, Sir Mortimer of Durand Line fame, rescued the article from oblivion and published it, along with the rest of his father's papers, in 1883. It prompted a writer in the *Edinburgh Review* to declare that Gardiner 'should long ago have enjoyed a world wide reputation'.

We know of others who listened to and were impressed by the Colonel's story. Lord Strathnairn when Commander-in-Chief in India, Charles Girdlestone and Le Poer Wynne, Cooper's successors in Srinagar, and perhaps Sir Richard Temple, who at one time or another occupied almost every top post in the Indian Civil Service. There was also Andrew Wilson, a globe-trotting writer for *Blackwoods Magazine*, who met him in 1873. Wilson was an urbane and accurate observer, less likely to be humbugged than most, and he had his reservations. Gardiner seemed to confuse hearsay with his own experience, 'but there is no doubt as to the general facts of his career'.

All these men dismissed the journals published by Edgeworth as garbled nonsense. They relied simply on what Gardiner told them and on their own judgement. And without exception they accepted his story. Indeed, considering the narrator's age, they were inclined to marvel that there were not more gaps and contradictions. All that was now needed to establish Gardiner's credibility once and for all was a full-length biography.

Fortuitously Cooper's draft and the notes with it were rediscovered. They were put in the hands of two 'very high authorities on Central Asia', both of whom died with the papers still in their possession. In the 1890s they reached a Major Hugh Pearse who referred them to Ney Elias, then unquestionably the greatest expert on the region and a man who had himself explored most of the Pamirs. But the jinx continued. Elias too died before completing his summary. Pearse, however, found his uncompleted notes on the manuscripts. In so far as they represent the only opinion of a nineteenth century geographer and explorer who was acquainted with the journals, Cooper's draft, Gardiner's unpublished notes and the accounts given by Durand and Wilson, they must be regarded as crucial.

Elias had written:

There appears to me to be good internal evidence that as regards the main routes he professes to have travelled, Gardiner's story is truthful. When he tells us that he visited the east coast of the Caspian, northern Persia, Herat, the Hazara country, even Khiva; that he spent some time in and about the district of Inderab, and afterwards passed through parts of Badakshan and Shignan, thence crossing the Pamirs into Eastern Turkestan, I see no reason to doubt him . . . The times were on the whole sufficiently favourable to render belief in the main features of his narrative possible; and it is in a sense the truth of the general narrative that enables us to excuse the untruth of many of the details. In other words had Gardiner not travelled over a great part of the ground he professes to describe, it would not have been possible for him to interpolate the doubtful portions of his story. He could not have known enough of the surrounding conditions or even the names of the places and tribes, nor have met with the people whose clumsy inventions he at times serves out to us. It is necessary, for instance, that a man who could never have read of the Pamir region should at least have visited that country or its neighbourhood before he could invent or repeat stories regarding Shakh Dara or the Yaman Yar, or be able to dictate the name of Shignan.

Using this verdict, Cooper's draft and all the other material, Pearse produced his standard work, *The Memoirs of Alexander Gardiner*, in 1898. It was twenty years after Gardiner's death and seventy after his journey through the Western Himalayas, far too late to create much of a stir. In the interim Turkestan and the Western Himalayas had been carved up between the great powers, there were patrols and boundary markers across the Pamirs, and in Chitral and Gilgit there were British garrisons. The authenticity, the very existence, of Alexander Gardiner was of purely academic interest. But at least his wandering spirit could now rest quietly in the little graveyard at Sialkot where he had been buried.

Not, however, for long. Pearse was wrong in thinking that only the most incredulous would now question Gardiner's story. The last blow was the cruellest of all. For in 1929 there appeared *European Adventurers of Northern India 1785-1849*, which not only reversed his verdict but damned Gardiner to perdition. 'Now', wrote the author, C. Grey, after citing the various authorities who had supported

Gardiner's tale, 'to impugn the veracity of any person is an unpleasant task, and especially so when that person was believed by men of unquestioned probity and high position to have been a worthy soldier and a truthful traveller.' It is 'only to be undertaken when one is certain of the real facts, has unimpeachable evidence to offer, no personal feeling and a conviction that by exposing an imposture he is rendering a service to posterity'. Grey knew what he was doing and enjoyed it. He had the evidence. With remorseless sarcasm and self-righteous zeal he proceeded to dissect limb by limb the Gardiner legend.

His suspicions had first been aroused by the similarity of incidents recorded by other Europeans in the Sikh service to those in Gardiner's tale. An Irishman called Rattray had also lent his diaries to Sir Alexander Burnes and lost them in the 1841 conflagration in Kabul. Leigh, a British deserter (alias Mohammed Khan) had also claimed extensive travels in Central Asia and Afghanistan, and so too had Honigberger, the German doctor who was dispenser of the Maharaja's brandy. Josiah Harlan was undoubtedly an American citizen but others, notably Charles Masson, who claimed to be Americans, were really British deserters. Gardiner, too, might be in this category. Grey was not prepared to credit him with even a fertile imagination. Every incident in his story could conceivably have been gleaned from the murky pasts of his companions at Lahore.

Suspicion became conviction when, working through the Punjab records, he came across the first of two damning pieces of evidence. This was an entry dated 15th December 1831 recording a report received in Ludhiana from the British informant at Lahore. It mentioned the arrival there of two Europeans, Messrs Khora* and Gardiner, both aged about 35. The date matches well with that given by Gardiner himself, though he never mentions a companion. Asked whence they came 'they answered that they were formerly serving in a ship of war, but not being satisfied with their position, quitted it and proceeded from Bombay to Peshawar, where Sultan Mohammed Khan [Dost Mohammed's brother] had entertained them on three rupees a day. They were with him six months, but having heard

---

* Khora, according to Grey, was really 'Kanara' or 'Kennedy', another Irish 'American'. He too stayed in the Sikh's service after Ranjit Singh's death and was seconded by the British in 1846. He died a hero's death at the outbreak of the second Sikh war in 1848.

of the liberality of His Highness [Ranjit Singh], they had applied for their discharge and come to Lahore.'

In other words, like so many adventurers, Gardiner was just a deserter from the East India Company's forces. Probably he was no more American than Masson and had never even been to Kabul, let alone crossed the deserts of Central Asia or the Himalayas.

Vanish [crows Grey] the highly accomplished parents of such singularly mixed breed and curious antecedents, and the long and tangled line of an adventurous life, 'one end commencing on the shores of Lake Superior and the other ending on the banks of the Indus'.

Gone are the many years of schooling in the Jesuit college, the long years of stirring travel, the high-minded Habib Ullah Khan, and those beautiful blue eyed, golden haired Kafir ladies, and the maiden captive who became Gardiner's wife and whose sad fate 'always brought tears to his eyes' . . . With them disappear . . . the companionship with princes, the command of armies and that adventurous travel and stirring adventure in lands 'where no European had ever set foot'.

Worse, in Grey's opinion, was to come. Gardiner was not only a liar but a brute. The other piece of evidence gleaned from the Punjab Records was that, during the anarchy that followed Ranjit Singh's death, Gardiner had acted as hatchet man for one of the aspirants to the throne. When no one else could be found for the role of torturer, it was he who had taken up the razor 'and with his own hands, in cold blood, without personal emnity of any sort' cut from a Brahmin, Jodha Ram, his right thumb, ears and nose. Every man has his price, and for this service he was accorded the equivalent of a Colonel's rank.

It needs to be emphasised that Gardiner was under orders and had little choice in the matter. Few survived those turbulent years in Lahore with their dignity, not to mention their lives, intact. But it was further recorded that, for this deed, Gardiner was expelled from Lahore by Sir Henry Lawrence when he arrived there as British Governor in 1846. There is no question that Gardiner did wield the razor, and many were understandably scandalised that any European officer could do such a thing. But the fact that expulsion merely took him to Ludhiana and safety, and that Lawrence afterwards spoke of him in the most generous terms, suggests that there were certainly extenuating circumstances.

Durand and Temple had made out the Colonel to be a paragon of virtue and 'a splendid example to the young man of today'. That he was no such thing is really beside the point. Grey had demolished his whole story. Legends based on it, and the reputation of the man who invented it, scarcely mattered any more. The question is whether Grey was right. Surely anyone with a spark of the romantic will be forgiven for trying to salvage the old boy's achievements, however unsavoury his character.

Grey's treatment leaves several questions unanswered. In the first place why, if he was really a deserter, was no action taken against him as soon as he reached British soil? Desertion was the sort of crime for which a man would be arraigned even decades later. It appears that during the anarchy in Lahore he supplied information to the British but there is no record, as there is with Masson who performed a similar service in Kabul, of a pardon being a quid pro quo.

His nationality, too, remains a puzzle. Von Hugel, who met him in Lahore in 1835 is the only one who states categorically that though he called himself an American he was in fact Irish. Where the Baron got this information is not revealed. Gardiner had indeed a brogue, but he admitted having spent five years in Ireland when he might easily have picked it up. As shown by Harlan's career, service in India did not preclude his being an American. It was improbable, but there is insufficient evidence to disprove it.

Grey is so sure that he is dealing with an 'imposture' that he can credit nothing of Gardiner's tale. 'All the claimed adventures will be found in books in our bibliography and such as are not therein in the accounts concerning Lee, Rattray, Jan Sahib, Vieskanawitch and other adventurers who served Ranjit Singh.' Unless Grey had a lot more information about these shadowy figures than he published, this simply is not true. There is nothing very original in his bibliography and, though it is possible that Gardiner pieced together from the pasts of these men an identikit life story for himself, nothing like enough is known about them to support Grey's assertion.

Of much more significance is the possibility, which escaped Grey, that the story which Gardiner gave to Ranjit Singh in Lahore was the false one; not that with which he regaled his audiences in Kashmir. The Maharaja wanted an army as disciplined as that of the British, and for this he wanted officers trained in European warfare and familiar with artillery. A wild adventurer from Central Asia with no understanding of drill or tactics was of no use to him. Such a man, to command a higher price, would naturally try to lay claim to some European service, and a supposed career in the navy would at least suggest that he knew something of gunnery. No native informant would be able to fault such a story; Gardiner admits that, knowing little about artillery, he had to bluff his way. Grey's evidence is unimpeachable all right, but, like that of Pearse, Durand and the rest, it too is based on Gardiner's word. In 1831 he had every reason to lie; in the 1870s, with one foot in the grave, very little.

Finally what of Gardiner's travels? Grey was no geographer and he completely ignored the testimony of Ney Elias, though he quoted in full the obsolete verdict of Yule. The 1860s and '70s were the golden age of travel in the Western Himalayas and explorers of the period invariably started from Srinagar. Had they, men who had seen or were soon to see the lands supposedly travelled by Gardiner, ever sought him out? It is a fairly obvious line of enquiry but one that neither Pearse nor Grey followed.

The answer, happily, is yes. Gordon, Forsyth, Hayward and Leitner all met Gardiner. Probably Shaw and possibly surveyors like Montgomerie and Godwin-Austen also knew him. Their references to him are scattered and far from conclusive. No one actually expresses an opinion one way or the other about his travels. But they all consulted him and there is no evidence that they considered him an imposter although they were men who knew the roads to Yarkand and Gilgit or who had visited the Pamirs.

Dr. Leitner merely quotes Gardiner as a connoisseur of feminine beauty throughout the Western Himalayas, and Forsyth refers to him as an authority on the mysterious mountains of Bolor.* Far more important is the testimony of Gordon and Hayward. Gordon first met Gardiner in 1852 when he watched him lead the troops at the Maharaja's weekly parade. Twenty years later, en route to the

---

* Bolor is one of those mysterious places like Atlantis and Eldorado which, though well attested by early travellers, including in this case Marco Polo, has never been identified. The explanation given by Gardiner to Forsyth, and much later by Younghusband to Pearse, is still the most probable. This assumes that Bolor is a mishearing of 'bala' meaning 'upper', i.e. the higher regions of any valley or tract. In other words it was not a place-name at all but a descriptive term. This, at least, explains why it appears to have been used of so many different places.

Pamirs, he again met him and had a long conversation with the old Colonel. It was before he had actually seen Yarkand or the Pamirs: but writing many years later he refers the reader to Pearse's *Memoirs of Alexander Gardiner* with obvious approval. Even after his travels he evidently had no doubts about Gardiner's story.

The same goes for Hayward. After his journey to Yarkand and Kashgar, he was actually concerned in trying to get the Colonel's story published. It could be that he was one of those 'very high authorities on Central Asia' who died whilst working on it, except that his death took place at a time when, according to Pearse, the papers were still with Cooper. Not only did he have no reservations about Gardiner but there is evidence to suggest that he was inspired by the old Colonel. A passion for disguise and travelling rough, a belief that the Chitral valley represented the easiest route from India to Central Asia and that the Pamirs should be approached via Gilgit, are all strongly indicative of Gardiner's teaching. This is purely conjecture but Hayward's references to him suggest that these two not dissimilar men were more than just acquaintances.

And there is one final piece of evidence. The authority in this case is a man who had no first-hand experience of anywhere beyond the British Indian frontier. He was neither a geographer nor an historian but a dogged, loyal and rather unimaginative civil servant. He had absolutely nothing in common with Gardiner, and if anyone was likely to be sceptical it was he.

Only three years before his death Gardiner, true to type, had been at the centre of a gun-running operation involving 20,000 muskets. These were supposed to be for the Amir of Kashgar, at the time regarded as a possible British ally in the Great Game. As it was, because of Gardiner's connection with the Kashmir army, the whole thing looked 'shaky' to the British authorities. The guns might never get further than Kashmir and in the Maharaja's hands could become a serious embarrassment. The transaction was therefore stopped. The letter dealing with the question went the rounds of the India Office in London. Rawlinson, who was a member of the India Council, records his doubts about Gardiner as a traveller but as usual avoids a definite opinion. Only one man, the diligent Under-Secretary Sir Owen Tudor Burne, is outspoken. He must have come across Gardiner when personal secretary to Lord Strathnairn in the early 1860s. The very existence of a connection between Gardiner and Burne is so improbable that his verdict, even if characteristically cautious, would be worth quoting. In fact it is anything but cautious. 'I know Gardiner well,' he notes in a neat hand in the margin. 'He is the most distinguished old man and has been through the whole of Central Asia experiencing the most extraordinary vicissitudes that ever befell the lot of one man.'

8. An Extraordinary Mission

Alexander Gardiner is really the antithesis of the nineteenth century explorer, a man from travel's other side. True or false, the important thing about his story is that it survives. It may stand for the unrecorded wanderings of many another vagrant and shadowy figure who perchance ventured into the Western Himalayas. In places like Australia the explorer could be confident of being the first outsider to set foot on a piece of virgin territory. This was not the case in Asia. Here there was always the possibility that, over the centuries of European contact with the East, some obscure wanderer had once strayed unknowingly along the same path. In Kabul there was a mysterious gravestone, seen by both Vigne and Charles Masson, which recorded a certain William Hicks who 'departed this lyfe' in 1666. In Kafiristan there was that legend of the two unknown Europeans, and in Kashmir Moorcroft had found a lone British deserter called Lyons and had enrolled him briefly in his private army. There was a White Russian called Danibeg who was supposed to have crossed from Leh to Yarkand at the end of the eighteenth century and there were all those mysterious adventurers, unearthed by Grey, who ended up in Ranjit Singh's army.

If a man like Joseph Wolff is a traveller but not an explorer, a further category must be found for Gardiner and those he represents. Travel to such men was a way of life; adventure its sole justification. Gardiner himself was not quite oblivious of geography. He may have possessed a compass and von Hugel found him touting a street plan of Kandahar. But maps were not part of his normal equipment and it was only after his travelling days were over that, in the hope of discovering just where his wanderings had taken him, he consulted one. At the time he can have had no idea if, when, or where he entered *terra incognita*. One desert, one range, was much like any other to a man preoccupied with staying alive. And although, after Gardiner, no European could be sure of being the first anywhere in the Western Himalayas, he himself would, at the time, have greeted with irreverent guffaws the suggestion that his desperate decampments merited the attentions of science.

Compare with such a man Lieutenant Alexander Burnes of the Bombay army. Gardiner's travels did nothing to advance a knowledge of the mountains; he scarcely knew of Moorcroft and he had no effect on the later course of exploration. Burnes, on the other hand, is a central figure in the story; and this in spite of the fact that he never actually set foot in the mountains. His career neatly links the achievements of all the early explorers. In his lifetime he was regarded as the greatest authority on the region and, as a result of his work, a whole new impetus was given to exploration. It was also under his auspices that another attempt was made on the Pamirs.

But first the man. It was not the least of Burnes's achievements that he won a degree of fame and recognition to which all subsequent travellers would aspire. In 1832, a few weeks after meeting Wolff in Kabul, he had reached Bukhara. He returned, aged twenty-seven, to the sort of reception every schoolboy dreams of. In a sense it had been easily won. He and Gerard had travelled light and in disguise, but beneath their ponderous turbans were hidden strings of gold ducats. More were sewn into their kummerbunds and stuffed into the soles of their sandals. Letters of credit for five thousand rupees were tied round their left arms and polyglot passports round their right. They were cherished employees of the Hon. Company and they enjoyed the best of both worlds; the freedom and anonymity of unofficial travellers together with the financial and political backing of accredited officers.

The risks involved were thus considerably reduced. But Burnes knew how to make the most of his success. He was a highly determined and ambitious young man, quick and canny as an east-coast Scotsman should be. 'Mercurial' is the adjective most often used of him. If Moorcroft's life is discerned in the slow uncertain light of a guttering candle and Gardiner's picked out with all the stagecraft of a *son et lumiere*, Burnes's is lit by the short-lived brilliance of a highly charged flashbulb. Likewise his character. A great talker, a good linguist and a shrewd diplomat, he had many of the qualities that showed. But later it would be asked whether he had the strong principles and staying power needed to go with them. He was also something of a scholar and, as might be expected of a distant descendant of Robert Burns,* an expressive writer. Anything his travels lost in reality they gained in the telling.

* Burns, Burnes and Burness are all variant spellings used by Robert's descendants.

From India he was sent straight to London to report in person to the Company's Board of Directors. *Travels into Bokhara* was published soon after he arrived and overnight 'Bokhara Burnes' became a legend. His observations were extensive but carefully checked, his assessments portentous but modest and discreet, and his narrative was nothing short of inspired. There was something of the wit of a Jacquemont, the sound common sense of a Vigne and the painstaking application of a Moorcroft. Yet none of these three had so far been published in England; almost all of what Burnes had to say was new. To the general public his book opened up a whole world of strange peoples and forgotten lands. Vicariously he invited the reader to share his adventures. They could feel the little thrill of excitement he experienced when a passer-by first asked him if he was heading for Bukhara. They could relish the anticipation of Hazara hospitality which reputedly included a bath and massage at the hands of the maiden of the house. And they could experience all the discomforts of native travel and the anxieties of disguise. It was the stuff of travellers' tales but Burnes made it respectable; he was not a crank but a highly authoritative and dedicated officer.

The book appeared in three beautifully bound, gilt-edged, gold-blocked volumes. The exceptional figure of £800 had been paid for the copyright and on the day of publication it sold nine hundred copies. The critics were unstinting. 'One of the most valuable books of travel that has ever appeared', thought the *Quarterly Review*; 'indispensable to statesmen, merchants, antiquarians and philosophers' chimed in the *Foreign and Quarterly*; while the *United Service Journal* was at a loss what to praise most, the statistical, geographical, geological, commercial, military or political information. For *Travels into Bokhara* was much more than just a travelogue; nothing less in fact than a scholarly and comprehensive account of the whole of Central Asia, the Western Himalayas included. Here could be found some of the first published reports of the source and course of the Upper Indus and of the Oxus, of Baltistan, Yarkand and the Pamirs. Political and scientific circles were as impressed as the general public. The young hero was introduced at Court and received the special thanks of the king. The prestigious Royal Asiatic Society elected him a member and conferred its diploma on him. The Royal Geographical Society elected him a fellow by acclamation and promptly awarded him their Gold Medal (until 1839 only one such award was made each year). And the Athenaeum Club co-opted him over the heads of 1,130 other candidates. The book was immediately translated into French and German, and Burnes dashed off to Paris to collect the plaudits of the French Royal Asiatic Society and the Silver Medal of the Geographical Society. He was 'living fast in every sense of the word' and loving it. Every reception afforded new introductions and more powerful connections. Burnes never underrated his prospects but, with an accent 'Scottish but not unpleasantly so' and a background respectable rather than influential, these were just what he needed to secure a brilliant future in his chosen field, the Indian political service.

The geographical societies would have done better to spare a plaudit for Dr. Gerard, who, though well past his prime, had joined Burnes at the latter's insistence to superintend the survey work of the expedition. From Persia Burnes had returned to India by sea while Gerard went back via Herat. He was already a sick man and took nearly eighteen months to reach Ludhiana. 'The trip killed him,' wrote one of his brothers, but he returned to their beloved retreat near Simla and there, on his deathbed, while Burnes was reaping the rewards in London, he completed the map of their travels. At a scale of five miles to the inch it measured ten feet by three and was the first attempt to link India to Central Asia by a connected series of observations.

Meanwhile Burnes was busy angling for an opportunity to return to Central Asia. Before dealing with the last stormy years of his short life, there are two aspects of the journey to Bukhara that deserve close examination. As has been said, excluding a visit to Simla, Burnes never set foot in the Western Himalayas. But the mountains intrigued him more even than the vast deserts of Central Asia. On his first visit to Lahore he had glimpsed the snowy tops on the horizon and experienced 'a nervous sensation of joy'. On his second visit, en route to Bukhara, he had actually sought Ranjit Singh's permission to cross into Kashmir; the Maharaja knew Burnes for what he was, a political agent, and duly withheld it. Both these visits to Lahore were in 1831, the same year that Jacquemont was in and out of the city on his way to Kashmir, and the same year that Gardiner finally arrived there. Early the following year Burnes met Wolff in Kabul and, on his triumphant return from Bukhara, it was he who unearthed Moorcroft's journals in Calcutta and took them back to London for publication.

He thus had his finger on the pulse of exploration into the Himalayas.

He may not have actually met Jacquemont, nor have yet got possession of any of Gardiner's journals, but he certainly used every scrap of information he had and used it to the best advantage. All his own material was hearsay. At Kunduz, where like Moorcroft he was hauled before Murad Beg, he met a party of merchants from Badakshan who were engaged in the tea trade between Yarkand and Bukhara. They provided a wealth of detail on the Pamirs and what lay beyond. Checked against what he knew of the earlier travellers and against the standard works of Elphinstone and Marco Polo, this provided a basis for an outline of the geography of the mountain region which vastly improved on anything then in print.

There were some glaring mistakes. MacCartney had given the Upper Indus three main branches. Burnes, eliminating the southern one, correctly reduced them to two but, ignoring Moorcroft, proclaimed the Shyok, and not the Ladakh branch, the main one. It was this error that sent Henderson and Vigne (until he exposed the mistake) on a wild goose chase after the supposed source of the Indus below the Karakoram pass. He was nearer a true idea of the positions of Chitral, Gilgit and Baltistan than anyone before Vigne but, contradicting even Jacquemont's evidence, made Skardu and Baltistan separate states. On the credit side Burnes included an account not only of the crossing of the Pamirs but also of the Leh–Yarkand route and it was from his book that the world first got an inkling of the mountain horrors that lay in store there. The Karakoram pass he reported as being high enough to cause severe altitude sickness and subject to storms so wild that they might delay travellers for a whole week. Beyond was a further pass of solid ice out of which steps had to be notched. On the basis of this information he doubted if the route would ever prove of much commercial significance.

However he dangled before the prospective traveller two tempting carrots, one of which was the city of Yarkand. Burnes was the first to sense something of the peculiar reputation of the place. With the merchants of Central Asia it was highly popular. Bukhara might be the noblest city, Merv the oldest and Samarkand the finest, but Yarkand was the naughtiest; it was their Paris. Western Turkestan was oppressively Mohammedan. The womenfolk kept within their windowless houses, and the visiting merchant was expected to reside in certain areas. Non-Mohammedans had to wear a distinctive rope girdle. If there had been any fleshpots, the outsider would not have had a taste of them. In Yarkand the native population was also Mohammedan, but it was ruled by the Chinese. Along with Kashgar, the other major city of Eastern Turkestan, it was the only place in the Chinese empire to which the Central Asian merchant was admitted. The Chinese vetted their visitors carefully and, once in, watched them closely. But this was not resented. Their rule was fair enough—and so too were the ladies of Yarkand. Mohammedans they might be, but they wore no veils and they hid from no man. On the contrary, from their high-heeled boots to their elaborate medieval head-dresses—all of which Burnes described in detail— they invited the very glances for which in Bukhara a man could be summarily arrested.

Some were prostitutes. There were streets of brothels and they were well patronised. But for the visiting merchant, who might spend months assembling his caravan, there were pleasanter arrangements. He got married. Four wives was the accepted maximum even in Yarkand, but a most convenient institution, what Burnes called a *nicka* marriage, limited the conjugal rights to a fixed period. It might be a year or two, or it might be just a week. And should even this prove too long, a cheap system of divorce offered an easy way out. By the time she was thirty an attractive Yarkandi girl might have had a hundred husbands. When, much later in the century, explorers and sportsmen would put into Yarkand to refit, their servants would immediately disperse into the bazaars and be married by nightfall. Winkling them out for the departure of the party was a delicate and trying business.

Burnes's other carrot, though every bit as intriguing, was more technical. In his description of the Pamirs he corroborated a reference by Marco Polo to the existence of a large lake in the centre of this barren chunk of mountains. He reported that it was called 'Surikol' and he learnt from men who had actually seen it that it was the source of the Oxus, the Jaxartes (Syr) and of a branch of the Indus. If this was the case, here indeed was a remarkable geographical phenomenon. Asia's three great classical rivers all flowing from the same icy tarn on the windswept 'Roof of the World'; it was a moving, almost allegorical notion. Lakes with more than one effluent are rare and the streams anyway are usually part of the same river system. But a lake with three effluents, each the origin of a separate mighty river system, was unheard of. Was such a thing possible? Did they ooze from a humble pool or gush from a noble sea? And where in that arid and elevated wilderness did the lake draw its inexhaustible supply? Until someone actually visited the scene and brought back an authentic account, Sir-i-kol was likely to prove as great a mystery as the snow-capped Mountains of the Moon round the equatorial source of the Nile.

So much for the geographical impetus provided by Burnes's trip to Bukhara. The other aspect of his travels which bears on the exploration of the Western Himalayas is political. With the appearance of Alexander Burnes, the Great Game, that depreciative catchphrase for the deadly serious business of Anglo-Russian rivalry in Central Asia, may be said to begin. In the 1820s the frontier of British India in the north-west lay from Simla along the line of the Sutlej to the deserts of Rajasthan, and then roughly followed the present Indo-Pakistan border to the Arabian Sea. Beyond it lay Kashmir and Ladakh in the north, Punjab and Afghanistan in the centre, and Sind and Baluchistan in the south. In a further arc beyond these lay Eastern Turkestan, the Central Asian Khanates of Khokand, Khiva and Bukhara, and Persia. Further still were the Aral, the Caspian and the Kazakh steppes—and only beyond these lay Russia. It seems incredible that anyone at the time can have contemplated the possibility of a Russian attack on India; nearly two thousand miles of desert, mountain and sea intervened. Yet at least two Tsars were thought to have planned such a move. Moorcroft had sniffed out Russian intrigues wherever he went, and in 1829 Lord Ellenborough, then President of the Government Board of Control for India, told the Duke of Wellington that invasion via Kabul, though not imminent, was decidedly possible. Burnes agreed wholeheartedly and on the way to Bukhara found further evidence of sinister Russian intentions. Only seven years later such mindless panic would greet the presence of a few Russian nationals in Afghanistan that a whole army would be despatched.

Without India the British colonies consisted of an inglorious rag-bag of semi-savage littorals. Yet India, the cornerstone of future empire, was now to be regarded as highly vulnerable. The threat came from the landlocked heart of Asia but British sovereignty rested on control of the seas. The enemy could assemble untold legions suitably equipped and acclimatised, within its own frontiers, whilst British reinforcements must undergo a four month sea voyage and then a long trek across the hot plains of India. On foreign terrain, or against a European power, neither the loyalty nor the fighting capacity of the Indian troops, who formed the bulk of the forces in India, could be taken for granted. The Russians, on the other hand, would be merely riding on the tide of history, sweeping into north-west India like countless previous invaders and no doubt aided and abetted by the Mohammedan powers of Central Asia who tradition-ally looked on Hindustan as a pampered maiden to be periodically ravished.

To those, like Burnes, who saw the threat in these terms the vital questions were where the main advances would take place and where they should be met. His antiquarian interests centred on the identi-fication of places mentioned in the histories of Alexander the Great and Timur. These men had led armies through much of the country a Russian force would have to cross. It was vital to establish their precise routes lest they be followed again. His geographical work was concentrated on assessing the feasibility of a modern army, complete with artillery, forcing the mountains or surviving in the deserts, and his political endeavours were directed towards gauging the strength and influencing the sympathies of those who ruled the intervening lands.

But there were others who actually welcomed the prospect of a Russian advance. Not into India, for this they considered beyond the wildest dreams of St. Petersberg, but as far as India. From Kashgar to Karachi the map was, in their opinion, disfigured by a patchwork of lawless, corrupt, slave-dealing states. These intervening lands, especially the Central Asian khanates, were in the last throes of decrepitude. Their rulers were cruel, bigoted pederasts and their power was founded on intrigue and terror. The very existence of such places was a challenge to all right-thinking men, and it was their instability which constituted the most serious threat to the prospect of amicable Anglo-Russian relations. The march of civilisation, as represented by Russian interests in the region, was inexorable. Britain had neither the might nor the right to oppose it and, should the eventual proximity of Russia ever tempt her to threaten India itself, that country would more easily be defended on its own frontier than in some fearful Armageddon in the unsurveyed depths of Asia.

This argument, besides dodging the vital question of just where the frontier of British India should lie, also overlooked what eventually became the over-riding consideration. This was the fear of contiguity. A common land frontier with a major European power was something for which Britannia was ill-equipped. Realists on both sides were inclined to pooh-pooh the idea of a Russian invasion. It was not just extravagant but unnecessary. As much could be gained if the Russians simply advanced as far as India. The Indian Army would then have to be augmented beyond anything that either India or the home country could afford. At the same time the co-operation of the subject peoples of the subcontinent would become highly doubtful. Imagine the situation in 1857 if the Indian mutineers could have looked to a sympathetic Goliath just across the Indus. In fact, in the long term, contiguity could prove more debilitating than a once-and-for-all trial of strength.

Here then were three ways of looking at Russian designs on India; the first military and alarmist, the second more ethical and consoling, and the third more realistic and political. They are just three shades from a whole spectrum of opinions. More people debated the Great Game than ever played it and there was more heated argument between the spectators than there were exchanges between the protagonists. Already it must be obvious that there was some confusion about the object of the exercise. Was it to prevent the Russians winning the hegemony of Central Asia, or was it to defend British possessions in India? The two were connected but they weren't necessarily the same thing. Secondly, who was playing the Game? In London there was the Foreign Office, the Company's Board of Directors and the Government Board of Control; in India there was the Governor-General in Calcutta and another semi-independent Presidency in Bombay. The authorities in India conducted their own relations with native states, both within the subcontinent and beyond it. Hence in Persia, for instance, both London and Calcutta were simultaneously represented. The Bombay presidency was nominally subordinate to the Governor-General but, in practice, had dealings with its immediate neighbours including trans-frontier states like Kutch and Sind. Yet a third area of debate concerned the tactics of the Game. Moorcroft had seen the first moves towards thwarting Russian designs as commercial and diplomatic, and Burnes agreed. But which were the markets that mattered, and who the princes to be supported? Strategically speaking, where lay 'the keys to India' and where the most defensible frontier? Was it on the Indus, the Khyber pass, the Afghan Hindu Kush or the Oxus? Within the Himalayas or beyond them? The subject is fascinating, inexhaustible and a trifle academic.

But to return to Burnes who, more than anyone, seemed to under-stand what the Game was all about. *Travels into Bokhara*, though discreetly written and carefully censored, brought the Russian threat into the open. His confidential reports went much further and convinced most of the players, though not perhaps Bentinck, that some initiative must now be taken. Burnes himself naturally favoured a 'forward' policy. He believed that Russian moves in Central Asia should be met, step for step, by British moves. At the moment Russia seemed busy with diplomatic and commercial operations in Persia and Bukhara. The British should follow suit, and the obvious target, because of its political instability and strategic importance, was Afghanistan. Burnes waited, confidently, for the order to return to Kabul.

Whilst in London he was offered the post of Secretary to the British Legation in Teheran. It meant promotion but not enough; he turned it down. 'I laugh at Persia and her politics; they are a bauble . . .' he told his brother. 'What are a colonelcy and a K.L.S. to me? I look far higher and shall either die or be so.' Vain words, but Burnes was like that. He stood a bare five feet nine and was of puny build. The portrait of him sitting cross-legged in Bokharan costume he thought needed a bit of touching up. His face looked 'so arch and cunning I shall be handed down to posterity as a real Tatar'. There is no fear of that. The wide beady eyes, the long bony nose and the smug cupid's bow mouth exude satisfaction rather than cunning. Fame at such an early age had somewhat turned his head.

He resumed his old post with the Bombay government and in 1836 his patience was rewarded. Lord Auckland took over from Bentinck as Governor-General, a more active foreign policy was adopted and Burnes, now a Captain, was duly directed to head an official mission into Afghanistan. Along with three other officers he left Bombay at midnight on the 26th November. Thus began what Charles Masson later called 'one of the most extraordinary missions ever sent forth by a government whether as to the manner in which it was conducted or as to the results'. Evidence supporting this verdict is not hard to come by. For a start the mission left by boat; a dhow took them to the mouth of the Indus and then, in a six-oared cutter, they rowed and were dragged upriver. It was not the quickest way of getting to Afghanistan; but it was essential if the mission was to retain an innocuous appearance. For these first tentative moves towards an assertion of British influence in Afghanistan were being disguised as a beneficent attempt to open a new trade route up the lower Indus.

Moorcroft had once mooted the idea of opening the Indus and Burnes, before his journey to Bukhara, had been the first European to sail up the river to the Punjab. But by 1836 he at least must have realised that here was something of a red herring. It is true that river navigation still had an important part to play in India. In fact in the short period between the arrival of the first steamship and the laying of the first railway track, it must have seemed crucial. But not surely on the Indus. In the first place the river, throughout its entire length, lay beyond the Company's frontier. All those through whose land it flowed, from the Amirs of Sind to Ranjit Singh in the Punjab, were strongly opposed to the scheme. Secondly, Burnes had already discovered that the river was not really suitable as a trade route. The shifting sandbanks made navigation of even the shallowest boats extremely hazardous, whilst the shortage of fuel on the semi-desert shores was an ill omen for the steamship. And thirdly, as Masson pointed out, it didn't need opening. It had never been closed. A certain amount of trade already passed up and down, customs duties were reasonable and there was nothing to prevent expansion if the demand were there.

Still, Burnes's mission was inaugurated as if it were a purely commercial exercise. He was to chart the river, to look for timber and coal on its shores, to select a suitable spot for an annual trade fair, and to proceed to Afghanistan simply in order to make commercial treaties with Kabul and Kandahar. It looks as if the Governor-General and his advisers had almost convinced themselves that the mission had no political objectives. It would not have done, certainly, for Ranjit Singh to have got wind of British approaches to his arch-enemy in Kabul, but it surely would have been prudent to have given Burnes some idea of how to deal with the sticky question of relations between the Sikhs and the Afghans. Or how to deal with the growing surge of rumours about Russian interest in the area. As it was, when a storm of events overtook the laden cutter in midstream, Burnes had to formulate his own.

The first rumble of thunder was distant but ominous. A Persian army was reported to be massing for an attack on Herat in western Afghanistan. Encouraged, partly financed and probably officered by Russians, it looked as if the hour of reckoning might already be nigh. Herat could not be expected to hold out for long and, as Burnes knew only too well from his strategic studies, once Herat fell the easiest of approaches to India, that via Kandahar, would lie wide open. Moreover there was every reason to fear that the rulers of Kandahar and Kabul would join the invader against their long-sworn enemy in Herat. Next, and considerably nearer to the mission, came a crash from the mouth of the Khyber pass. At Jamrud near Peshawar the Sikhs and Afghans fought a full-scale battle. Dost Mohammed of Kabul had never resigned himself to the loss of Peshawar and, though his troops failed to reach it, or indeed to win a convincing victory, he was sufficiently encouraged by the result to think that Ranjit Singh might yet give it up. Burnes had no instructions as to how to cope with this quarrel but, since both Sikhs and Afghans might prove vital allies in the trial of strength that loomed ahead, it was imperative that he help find a quick solution.

During the hot months of 1837, as the mission drew nearer to Kabul, Burnes began to sense that history was in the making, and that he alone was in a unique position to influence it. He had come to look after trade and surveys, he told his brother, to gauge the state of affairs and to see what could be done thereafter. 'But the hereafter has already arrived . . . I was ordered to pause but forward is my motto; forward to the scene of carnage where, instead of embarrassing my government, I feel myself in a situation to do good.' He was throbbing with excitement. News of Russian overtures to Kabul, he told Masson, 'convince me a stirring time for *political* action has arrived'. And when rumours of similar approaches were also reported from Kandahar, 'Why zounds, this is carrying the fire to our door with a vengeance.'

The mission reached Kabul in September 1837. Burnes was already on good terms with Dost Mohammed. In 1832 en route to Bukhara he had been offered that old and deceptive carrot, the command of his army, and had left Kabul with sincere regrets. He genuinely liked the people and was now more welcome than ever. Preceded by a friendly letter from the new Governor-General, his arrival at such an opportune moment was presumed to mean that the British were prepared to put pressure on Ranjit Singh for the restitution of Peshawar. Like most native princes, Dost Mohammed had no interest in a commercial treaty. As he understood it, this was just an odd European euphemism for a political alliance; but even an alliance he would willingly concede if it meant the return of Peshawar. Burnes pondered the situation. Herat was now under siege, Kabul and Kandahar could turn to the Persians at any moment, and it was too late to seek new instructions from Calcutta. He decided that the only way to retain control of matters was by a sympathetic consideration of the Peshawar question, and he opened the ill-fated negotiations on that basis.

9. The Source of the Oxus

While Burnes juggled with the fate of nations in Kabul, another young man, in circumstances no less unexpected, was on his way to achieving the one solid success of the whole 'extraordinary' mission. Burnes had with him three British companions, a doctor and political adviser, Percival Lord, a soldier, Lieutenant Leech of the Bombay Infantry, and a sailor, Lieutenant Wood of the Indian Navy. All possessed some knowledge of surveying and on arrival in Kabul they were sent into the Kohistan, Gardiner's old stamping ground; they were to continue the study of the Hindu Kush passes. But almost immediately they were recalled; Leech was packed off to watch the situation in Kandahar, and Lord, most unexpectedly, found himself invited to Kunduz.

From Kunduz Murad Beg still ruled his insalubrious foothills and fens along the banks of the Oxus. But, with old age, his character must have mellowed. He had a brother, Mohammed Beg, who was slowly going blind. The brother had remained a loyal and suitably cruel adherent throughout the chief's long, rapacious career. Such fraternal devotion was far from normal in the context of Afghan politics. It deserved acknowledgement, though Murad Beg in his younger days would scarcely have recognised the obligation. Now, however, he was prepared even to help solve the mystery of Moorcroft if Burnes would send the 'feringhi hakim' to treat his brother.

It was the first piece of good news the mission had had. Dr. Gerard on his return from Bukhara had been rebuffed by the old chief and so, too, had Vigne during his Afghan year of 1836. Now there was at last a chance of opening relations with Kunduz and treating with Murad Beg from a position stronger than that of a helpless captive. After exacting guarantees of safe conduct Lord left for Kunduz on November 5th 1837. It seems to have been something of an afterthought to send Wood along too. In 1835 he had been the first man to take a steamer up the lower Indus and, on a mission that was supposed to be charting the river and opening it as a trade route, there was good reason for his inclusion. But in Central Asia he must have begun to feel rather redundant; there was not much work for a marine surveyor. He had exhausted the possibilities of the Kabul river and, since the Oxus was the next river of any significance between Russia and India, Burnes decided to send him on to Kunduz.

On the previous journey Burnes had been impressed by the formidable character of the Afghan Hindu Kush. Crossing just after Wolff had performed his naked flit he had found the snow still deep, the cold intense and the track, that via Bamian which was reputedly the easiest, far from gentle.* It seemed unlikely that an invading army would attempt such a crossing and, unless there were easier passes elsewhere, he suspected that an advance must be made either south of the mountains via Herat and Kandahar or, perhaps, north along the line of the Oxus towards Chitral and Kashmir. He had already sent a man up through Chitral to assess this approach. He was convinced the Oxus was navigable as far as Kunduz, but beyond there it was an unknown quantity. If at all possible, Wood was to investigate the subject.

Nothing was said about the Pamirs or Sir-i-kol. No doubt they had discussed the mysterious lake but Burnes was currently more preoccupied with the fate of nations than with geographical oddities. Only Wood, who knew and loved his Marco Polo, had begun to discern something like the hand of destiny beckoning him towards the Pamirs. There is something so improbable about the first Briton to stand on the Roof of the World being a naval lieutenant that one can understand a certain precognition as he was sucked by circumstance further and further into the heart of Asia. A month later he recorded in his journal how 'the great object of my thoughts by day and my dreams by night had for some time past been the discovery of the sources of the Oxus'.

The first attempt to cross the Hindu Kush failed and it was not till December 4th that Lord and Wood finally reached Kunduz. The dates were now becoming crucial. If Wood actually got permission to realise his dream and ascend the Oxus, it was bound to be a wintry journey. The cold would be intense, but this was no problem beside the delays that might be caused by blizzards. Worst of all, there was the risk of being caught by the first thaw when the snow would become soft and all possibility of movement have to be abandoned for weeks.

Mohammed Beg's trouble was diagnosed as ophthalmia. One eye

* Dr. Gerard disagreed about the Afghan Hindu Kush. After his experience of the Great Himalaya east of Simla, he regarded the passes to Bamian as child's play and quite feasible as invasion routes.

was already blind and the other failing fast. Probably Lord realised that there was nothing he could do about it. But he wanted to stay on for the winter and to make the most of Murad Beg's goodwill. He therefore embarked on a course of treatment and, since it might take some time, suggested to his host that Wood be allowed to occupy himself with a trip up the Oxus. He may even have inferred that favourable consideration of this request would assist the success of his treatment; the Uzbeks had a high, even superstitions, regard for European medicine. But equally Murad Beg may have required little persuasion. For all his faults his idea of dominion had none of the jealous political awareness of Ranjit Singh's. Wood describes him as able and despotic but still no more than 'the head of an organised banditti'. He had little to fear from his neighbours and no comprehension of the brewing power struggle beyond. Wood's plan must have sounded like an eccentric but harmless exercise and he readily agreed.

> Monday the 11th December was fortunately a market day in Kunduz; so that the articles required for our expedition were at once obtained and lest Murad Beg might recall the permission he had given, we started the same evening for Badakshan and the Oxus. We adopted the costume of the country, as a measure calculated to smooth our intercourse with a strange people, and we had little baggage to excite cupidity or suspicion. Coarse clothes to barter for food with the inhabitants of the mountains, was our stock in trade; and my chronometers and other instruments the only articles of value I took with me. Dr. Lord accompanied us for the first few miles, and parted from us with cordial wishes for the success of our expedition.

Thus with bland simplicity begins the story of John Wood's lone journey to the source of the Oxus. It was thrown together from the rough notes of his journal, 'filed at the ends and twisted into narrative form'. That was about the strength of his prose; neat and manageable but not exactly inspiring. Wood too was like that, a quiet, capable but unobtrusive man. Sir Henry Rawlinson's obituary notice of him singles out as typical 'an extremely retiring disposition, in him amounting to second nature, which often prevented his coming before the world so prominently as his friends desired'. Compared to Burnes he certainly betrayed an 'invincible modesty'. But one must beware of such epithets from other explorers. Most were in no position to recognise modesty when they saw it. Basking in the glare of publicity, they ogled acclaim and revelled in controversy to an extent that is now inconceivable. Could anything be more undignified than Wolff and Burnes, the only men alive who had visited Bukhara, slogging it out in the Calcutta newspapers over whether or not Wolff had travelled in disguise?

John Wood was not like this. He shunned publicity. Home was not mingling with the mighty in London but a quiet corner of Perthshire or Argyll. This, in the eyes of a man like Rawlinson, made him a recluse, and his career does confirm the impression of a man running away from his achievements. Yet no one who relished obscurity was likely to find himself unexpectedly projected into the heart of Asia. Ambition and vision were as essential as stamina and determination. Wood had all of these; he was also a typical sailor. Long solitary hours on the bridge of some minuscule puffer had given him an introspective frame of mind. He was a bit of a dreamer, but a practical dreamer. Rather than argue his ideas he liked to test them, and if this absented him from the halls of fame so much the better.

The preparations for the journey were meagre. An elaborate expedition was out of the question and the only chance of success lay in a wild dash before the winter started to lose its grip. No tents were taken and no food or special equipment. They would travel fast and light. There was virtually no baggage and every man—Wood had seven followers—was mounted. They relied entirely on such hospitality as the authority of Murad Beg's representatives might provide. In the best Burnes tradition Wood wore Uzbek clothes, but it was not a question of disguise. A tall thin man, he stood out like a sore thumb amongst the diminutive Mongol races of the mountains. Nor were political difficulties anticipated. It was just the most practical way to travel.

As far as Jerm, then the principal town of Badakshan, all went to plan. Riding up to forty miles a day they covered the distance in a week. The countryside was as devastated as it had been in Gardiner's day. Once upon a time Badakshan had been a land of milk and honey, famed by the Persian poets for its fruits, its jewels and its climate. Now the orchards were overgrown, the mines unworked, the villages in ruins and the terraced fields barely distinguishable. The population had been slain in their thousands and the survivors Murad Beg had deported to Kunduz, 'a place', according to Wood, 'only fit to be the residence of aquatic birds'. And of course it was mid-winter. The rain and mists of Kunduz gave way to the snows and piercing winds of the mountain districts. Approaching Jerm it was 25 degrees below freezing and they were only at 6,000 feet, a fraction of the altitude of the Pamirs.

Wood, however, was in the best of spirits, as delighted to be rid of the mud and squalor of Kunduz as of the political activities of Lord and Burnes. He preferred the simple, independent life. At night he invariably found 'a snug berth' in one of the few inhabited villages. Toasting his toes before an open fire dexterously tended by some demure Tajik housewife, he would sip a cup of thick, salted tea and enjoy that most delicious of travellers' sensations, '. . . a glow at the heart that cannot be described. A calmness of spirit, a willingness to be satisfied and pleased with everything around me, and a desire that others should be as happy as myself.' It is no ecstasy, but a quiet warm joy occasioned more by what may be simmering for supper on the stone hearth than by any lofty sense of achievement.

He reached Jerm on December 18th but not till January 30th did he resume the journey. For six weeks he was stuck. The blizzards of mid-winter had arrived sooner than usual. Ahead the road to Wakhan and the Pamirs was blocked. He was told that even a short excursion down the Oxus to the ruby mines, the same as those visited by Gardiner, was out of the question. While the snow outside accumulated to a depth of several feet Wood sat by the fire 'in a brown study' and rued the missed gaieties of Christmas and Hogmanay back home in Scotland. To stave off boredom he studied the Badakshi language and entered enthusiastically into the social life of the town. There was much to admire in the people. Rich and poor lived on terms of social equality and even children might attend the councils of state. Wood was not the man to try and impress them with 'chemical experiments' or the wonders of European technology. In his open-minded appreciation of a totally alien yet practical existence there is a hint of the dissatisfaction with society at home that came later to dictate his career.

On January 30th a party from Shignan arrived in Jerm. They had found the Oxus frozen solid and had travelled on it past the ruby mines to Wakhan and thence to Jerm. So the road was now open, and Wood could at least reach the river and explore it down to the mines, even if the onward journey to its source was still doubtful. He left for Wakhan immediately.

The temperature was now dropping to six degrees Fahrenheit, the wind worse than ever and blizzards frequent. Climbing the 10,000-

foot pass from Zebak to Ishkashim they met a solitary traveller dressed in the skin of a horse. He had come up the Oxus from Darwaz beyond Shignan. The ice, he said, was already breaking up and the steep rocky walls which contained the river were even more treacherous. In the intense cold he owed his survival entirely to his steed; no longer able to ride him, he had sadly elected to wear him. Hard on his heels came another party who had also ascended the Oxus as far as Wakhan. They confirmed the bad news that winter was too far advanced. One man had lost not just a finger but a whole arm from frostbite. But much worse was their confirmation that the river below Wakhan was now so broken up that they had had to travel by a path along the steep banks. It was treacherous enough, but a greater danger was from avalanches. In one such they had lost half their party, mules, men and baggage all swept without trace into the river.

Between the Pamirs and Kunduz the Oxus wanders from its east–west course into a long loop round Badakshan. Thus far Wood had been cutting across the open end of this loop and so had still not actually seen the river on which his hopes were pinned. Now, as he topped the pass above Ishkashim, he looked down for the first time on the Oxus valley. It was not an encouraging sight. The valley here is still inhabited, comprising the small and very poor principality of Wakhan. But from the pass there was no evidence of life, just an utterly featureless snowscape, flanked to the south by the peaks of the Hindu Kush and to the north by the swelling slopes of the Pamirs. Of the river there was no sign. Like everything else it was buried beneath the snow.

Gardiner mentions getting his horse across the Oxus by lashing blocks of ice together and covering them with straw; an unlikely tale until one compares it with Wood's experience. He crossed at Ishkashim on bridges of frozen snow and, in spite of the gruesome accounts he had heard, attempted to make an excursion downstream to the ruby mines. These mines, together with some lapis lazuli workings near Jerm which he had already visited, constituted the great attraction of the region. Badakshan had once been famous for its jewels, and the mines were one of the main reasons for Murad Beg's conquest of it; they were now boarded up, largely because of his fear of the gems being stolen. Wood's interest was simply to try and assess the value of the country's one known commodity. In the event it was impossible. The river beneath its sheet of ice and snow had risen and was now covered with vast slabs of ice which moved and tipped pre-

9 Colonel Alexander Gardiner

10 Maharaja Ranjit Singh, from a portrait by G. T. Vigne

11 Dost Mohammed of Kabul, f
a portrait by G. T. Vigne

12 Sir Alexander Burnes

cariously. With the utmost reluctance he turned back to Ishkashim.

This left just the final and principal object of the journey, to follow the river upstream to its source. For the fifty-odd miles through the bleak Wakhan valley, all went well. Wood was getting used to the cold and the higher they proceeded up the river, the firmer became the ice. But now new problems arose. At Qala Panja he found that the river divided into two, and he had to decide which was the main feeder. It was a difficult decision. The local people insisted that the more northerly branch was the Oxus. In summer it was by far the larger and it commenced from the lake called Sir-i-kol. This tied in well with the evidence gathered by Marco Polo and Burnes. But Wood was still unsure, and with measuring ropes and a thermometer clambered down to the confluence.

Had he, as he slithered across the ice and crouched among the boulders, had any idea of the significance that would attach to his decision or of the controversy to which it would give rise, he must still have reached the same conclusion. He admits that the southerly branch, the Sarhad, looked the larger. But then it was free of ice and flowing in one channel. The Sir-i-kol branch was split into several channels, all frozen over, so that its volume was impossible to judge. Beneath the ice its current did seem to be the faster of the two and its temperature much the lower. Wood thought this indicated a loftier source and, taking all in all, plumped for the Sir-i-kol.

There is no single criterion by which geographers resolve a quandary like this. Even the longest branch, something which, given the season, Wood could hardly be expected to investigate, is not necessarily accepted as the parent river. Volume of flow, direction of flow, speed of current, altitude of origin, and local and historical tradition are all relevant factors. Wood was wrongly informed that the Sarhad came from above Chitral, that is from the Hindu Kush, and since it was generally agreed that the Oxus rose in the Pamirs this too influenced his decision. He acknowledged that he was contradicting the evidence gathered by MacCartney, which seemed to favour this branch, and he admitted that it appeared to have more feeders than the Sir-i-kol. In fact his decision was as qualified and cautious as one would expect from a man of 'invincible modesty'. But given the circumstances and the interest attaching to the lake of 'Surikol' it is hard to see how, even if the northerly branch had obviously been the junior, he could have foregone the chance of penetrating the Pamirs and solving the mystery of their famous lake.

The other problem was to find guides and an escort for the last fearsome haul. His Afghan and Uzbek companions were unanimously opposed to further progress and seemed to be in collusion with the already uncooperative Mir of Wakhan. Wood countered this by appealing for help to the very people whose predatory habits necessitated the escort, the Kirghiz of the Pamirs. By a lucky coincidence, a large horde of these nomadic shepherds had, for the winter of 1837–38, come down to Wakhan. Wood had already noticed their sturdy yaks, 'like giant Newfoundland dogs', and their hairy Bactrian camels; he had even stayed in one of their rounded black tents. They were scarcely as untrustworthy as the Wakhis made out, and his confidence in them was rewarded by the appearance of a party of mounted men willing to chance a winter foray up to Sir-i-kol.

Bundled up like Eskimos, they set forth in single file on the final stage. The country beyond Qala Panja was uninhabited and treeless. They carried eight days' food and some fuel but they were still without tents. On the first night, at an altitude of about 12,000 feet, the mercury fell to six degrees and disappeared into the bulb at the bottom. They sought shelter from the wind among the baggage and horses, but still three men were so badly frostbitten that they had to be sent back. All next day and the following one they plodded on up the frozen river. For fuel they used dung, dug from the sites of Kirghiz summer encampments. Their only shelter came from the walls of snow, three feet deep, resulting from these excavations.

At the second 'camp' more men and provisions were left behind and a further mutinous group abandoned at the third. They were now at 14,000 feet and thought to be only twenty miles from the lake. Wood, if needs be, was prepared to go it alone but luckily, for the worst was still to come, five of his men remained loyal. Thus far they had been following the tracks of a returning party of Kirghiz. These now abruptly ceased, and on the fourth day they slowed to a rate of five hundred yards in two hours as they fought their way through fields of deep snow. 'Each individual of the party by turns took the lead, and forced his horse to struggle onwards till exhaustion brought it down in the snow where it was allowed to lie and recruit whilst the next was urged forward.' Returning to the frozen surface of the river they made better progress but the ice soon became too brittle and weak to be trusted. A mule crashing through into the icy waters was taken as a sure sign that their goal was at last nigh.

After quitting the surface of the river we travelled about an hour along its right bank, and then ascended a low hill, which apparently bounded the valley to the eastward; on surmounting this, at five o'clock on the afternoon of the 19th February 1838, we stood, to use a native expression, upon the *Bam-i-Duniah* or Roof of the World, while before us lay stretched a noble but frozen sheet of water, from whose western end issued the infant river of the Oxus. On three sides it is bordered by swelling hills, about 500 feet high, whilst along its southern bank they rise into mountains 3,500 feet above the lake or 19,000 above the sea, and covered with perpetual snow, from which never failing source the lake is supplied . . . As I had the good fortune to be the first European who in later times had succeeded in reaching the sources of this river, and as, shortly before setting out on my journey, we had received the news of Her Gracious Majesty's accession to the throne, I was much tempted to apply the name of Victoria to this, if I may so term it, newly re-discovered lake; but on considering that by thus introducing a new name, however honoured, into our maps, great confusion in geography might arise, I deemed it better to retain the name of Sir-i-kol, the appellation given to it by our guides.

Wood's decision not to call the lake Victoria is typical of the man. It seems equally inevitable that his cautious reasoning was later ignored and, before Speke set eyes on Lake Nyanza, Sir-i-kol was already known as Lake Victoria. As to the other rivers which supposedly flowed from it, Burnes had been misled. There was only one effluent, the Oxus, which in mid-winter was just five yards wide and ankle deep.

The depth of the lake itself posed something of a problem. From Qala Panja Wood had brought a line of a hundred fathoms and with this, early next morning, he sallied forth onto the ice. They shovelled away the snow and then with pickaxes started to chip into the ice. Each man had a few strokes and then collapsed into the snow exhausted. The difficulty of breathing at high altitudes, what Trebeck had called 'a frequent inclination, and at the same time a sense of inability, to sigh', was painfully evident. Two and a half feet down, a final and imprudent swing of the pick produced a hole too small to take the lead and a jet of water which drenched them all. Puffing with exhaustion and festooned with icicles from the soaking, they tried again in another spot. As they approached the danger zone four men stood round the hole with a rock poised above their heads. The pick shattered the final layer, at great speed the rock followed it and at last there was a hole of serviceable size. No one was soaked and Wood was soon paying out the line in the best naval fashion. But not for long. At just nine feet he struck the oozy, weed-strewn bottom.

If the lake itself was a bit of an anticlimax, not so the Pamirs. Curiously Wood continues to describe the region as an elevated plateau although, on his own evidence, there were mountains of 19,000 feet. He confirmed Marco Polo's assertion of the ground apparently falling away in all directions, and he saw the region as the focal point in both the hydrography and orography of the continent. The Syr did not rise in Sir-i-kol but, along with a branch of the Tarim river flowing east towards China and the Chitral river flowing south to the Indus, he imagined that it drained the Pamirs. In fact he was wrong about both the Syr and the Chitral, but he was on safer ground in asserting that the mountains of Asia seemed to radiate from the Pamirs. Not only the ranges of the Western Himalayas but also the Alai and Tian Shan connect to them.

Wood makes light of the cold though, like many a polar traveller, he found the desolation unbearable.

> Wherever the eye fell one dazzling sheet of snow carpeted the ground, while the sky everywhere was of a dark and angry hue. Clouds would have been a relief to the eye; but they were wanting. Not a breath moved along the surface of the lake; not a beast, not even a bird, was visible. The sound of the human voice would have been music to the ear, but no one at this inhospitable season thinks of invading these gelid domains. Silence reigned around—silence so profound that it oppressed the heart and, as I contemplated the hoary summits of the everlasting mountains, where human foot had never trod, and where lay piled the snow of ages, my own dear country and all the social blessings it contains passed across my mind with a vividness of recollection that I had never felt before. It is all very well for men in crowded cities to be disgusted with the world and to talk of the delights of solitude. Let them pass but one twenty-four hours on the banks of the Sir-i-kol and it will do more to make them contented with their lot than a thousand arguments.

Wood rarely writes from the heart like this. He was not so much interested in reassuring the 'men in crowded cities' as in recording his own love–hate feelings about isolation. He goes on to protest that man's proper place is in society but his own career scarcely lived up to this dictum. From Sir-i-kol he turned back to Qala Panja, back through Wakhan and Badakshan to Kunduz and then back with Lord to Kabul, Ludhiana and Bombay. Soon after, he left the Indian Navy and returned to Scotland. In 1841 the Royal Geographical Society awarded him their Patron's Gold Medal. Wood did not even bother to go to London to receive it, and a few weeks later he emigrated with his family to New Zealand, a place then boasting little more in the way of society than the Pamirs. It was not this, but the misrepresentations put out by the colonial developers that turned him against the place, and he was soon back in Britain. In 1849 Lord Napier, one of those British officers who had tended the disaster-prone Wolff and who was now appointed Commander-in-Chief in India, requested his services. The request was vetoed by the Company's Board of Directors. In disgust Wood emigrated again, this time to Australia. He did eventually return to India but not in the service of the government. Preferring 'rather to wear out than to rust out', he passed his last years back on the Indus and commanded a steam flotilla, that by then sailed regularly up through Sind, until his death in 1871. To discover why a career begun so brilliantly ended in comparative obscurity, we must return to Burnes and Lord and the tragic events of 1838–41 in Afghanistan.

*   *   *

Lord had spent the winter in Kunduz rounding up some of Moorcroft's scattered possessions and, under cover of this harmless occupation, organising a network of British informants along the middle Oxus. In spring, well pleased with their work, he and Wood recrossed the Afghan Hindu Kush to Kabul. There, all was confusion. Burnes had already retired. His negotiations with Dost Mohammed had collapsed and a Russian envoy had taken his place. The blame for all this, and for the subsequent hostilities, rests almost entirely on the shoulders of Lord Auckland and his advisers. Burnes's behaviour is open to criticism. Masson found him too credulous of bazaar gossip, too obsequious towards Dost Mohammed and far too alarmist. He certainly exceeded his instructions and completely misread the mood of his superiors. He also scandalised those who admired him most by maintaining a harem of 'black-eyed damsels'; in the jargon of the day, it betrayed a weakness of character.

But had he been given clearer instructions about how far he might go in accommodating Dost Mohammed, indeed had his superiors been decided on this point, all might have been well. Or had Wade in Ludhiana not conceived a jealous mistrust of his activities and wilfully misrepresented his reports to the government, his predicament might at least have been understood. Burnes argued that the Sikhs must be persuaded to accommodate Dost Mohammed's claim to Peshawar. This was the price demanded by the Dost for repudiating Russian approaches and, since these appeared to add cogency to the argument, Burnes tended to inflate them. But Wade, who was a member of the Bengal service, grew increasingly suspicious of Burnes and his Bombay men. Their schemes seemed to threaten his long-standing relationship with the Sikhs, and he insisted that Ranjit Singh could under no circumstances be induced to surrender Peshawar. The endless reports from Masson and Burnes about the likelihood of Russian intervention in Kabul he represented to the Governor-General as reason, not for befriending Dost Mohammed, but for abandoning him and his already tainted country.

Meanwhile Herat was still under siege. To many in India it seemed that British rule in Asia hung in the balance. If Herat fell, if Kandahar, already wavering, went over to the Perso-Russian camp and if Kabul followed suit, then the gates of India would have fallen. To men like Wade and Auckland's secretary, MacNaghten, it was simply a question of whether this was the time to trust a new ally at the expense of an old one; better, surely, to consolidate the old alliance and help Ranjit Singh, or a joint protégé, on to the throne of Kabul.

On Christmas Eve 1837, while Wood was stuck in Badakshan, Lord in Kunduz and Auckland being swayed to and fro by his advisers in Calcutta, a Captain Vitkievich, with a letter three feet long offering the Dost Russian cash to attack the Sikhs, arrived in Kabul. Burnes was non-plussed. But here was evidence incontrovertible of the gravity of affairs. He sent an express back to the government and confidently awaited the adoption of his plans which to him seemed more certain and urgent than ever. It was not forthcoming. On the contrary, his negotiations with both Kabul and Kandahar were repudiated. The game, in his own words, was up. Vitkievich had an open field and Burnes, unable to wait for his companions, slunk away from what was to have been the scene of his greatest triumph.

He fell back on Peshawar and then on Lahore where MacNaghten was already planning the next move in the fatal scenario. Dost Mohammed had entertained a Russian envoy and rejected a British mission. That he had been forced into this position was ignored. All that mattered was that now he was obviously an unsuitable ally and, in the face of the supposed threat from Herat, must needs be quickly replaced by someone more amenable. Shah Shuja, an ex-king of Kabul and currently an Anglo-Sikh pensioner, was the obvious choice. A tubby little man as inoffensive as he was ineffectual, it is hard to see how he could ever have held the Afghans together. Burnes, in May 1838, wrote that he at least would never lead the ex-king against his old friend, Dost Mohammed. And he didn't; not actually lead him, that is. But before the year was out, promoted to Colonel, elevated to a knighthood, still only thirty-three and still angling for the job of Resident in Kabul, he was smoothing the way for the Army of the Indus which was to install Shah Shuja on the throne of Kabul.

Such resilience, in his own view, scarcely smacked of compromise. He claimed to have espoused Shah Shuja's cause in the interests of his country, at great personal sacrifice and simply in order to help the government out of an impasse of their own creating. It was also suggested that he only did so because Auckland's alternative, to let Ranjit Singh have Kabul, was infinitely worse. But this was protesting too much. The fact was that Burnes had performed an about face. He was now deposing the man he regarded as the ablest of rulers and the closest of friends. Not surprisingly his conscience pricked.

Lord followed Burnes's lead and only John Wood, who can hardly have had much to do with the Dost, baulked. To his mind it was a sell-out. Though at the height of his fame he refused the bitter pill and, with quiet dignity, resigned. If the Board of Directors' behaviour ten years later is anything to go by, he was never forgiven. Even on their Commander-in-Chief's recommendation they refused to have him back in their service. One wonders whether that 'retiring disposition' was not more the product of circumstance than of character.

Wood's only reward for this gesture was survival. Lord, Burnes and the Army of the Indus never returned. Though the Persians retreated from Herat and with them the Russian threat, the Army of the Indus continued its fatal progress to Kabul. In stony silence, Shah Shuja was duly installed on the throne. The army stayed on, outliving the welcome it had never had, and in late 1840, near Gardiner's old *castello* at Parwan, the first serious reverse occurred. An order was given for a British force to charge the massed ranks of Dost Mohammed and his son. Not a man stirred. Only the officers,

Lord amongst them, sallied forth and were duly massacred. The following September a tribal rising trapped another British column in Jalalabad and closed the most direct bolt hole back to India. Burnes was still in Kabul, bitterly disillusioned by the squabbles within the occupying army but biding his time till MacNaghten left him in sole command. He was trying to stay aloof and he set up home in the heart of the city, surrounded by oriental splendour and with an overflowing seraglio. 'I grew very tired of praise,' he wrote, 'but I suppose I shall get tired of censure in time.' The final censure was about to come from the least expected quarter. He thought he knew the Afghans better than anyone. Long ago, in the pages of *Travels into Bokhara*, he had declared that no race was less capable of intrigue. Now he sensed danger, but not so much for himself as for the whole invading force. He continued to live miles from the British garrison and to flaunt his adoption of the courtly ways of Kabul. On November 2nd 1841 the end came swiftly. The 'guileless' Afghans appeared before his house in strength, fired the building and knifed to death its inmates as they attempted to escape. No one survived.

Thus ended the mercurial career of Colonel Sir Alexander Burnes. For ten short and controversial years his every move had been made in the full glare of public attention. If he died with his ambition somewhat blunted by cynicism it was hardly surprising. Celebrity, censure and then honours, all had come for the wrong reasons. Even his death had its irony. Not only did he trust the Afghans, but he above all others had their best interests at heart. Yet the Afghans selected Burnes as the first scapegoat.

His death was the signal for the general uprising. From the whole Kabul garrison only one man reached Jalalabad alive. MacNaghten, Shah Shuja and all the other principals were massacred. The story has been often told, but it was indeed the worst ever defeat suffered by the British army in Asia, one not equalled till the fall of Singapore exactly a hundred years later.

In time the fragments of Anglo-Afghan policy were picked up. Honour was restored by a second expedition which secured the release of prisoners and meted out reprisals. Dost Mohammed was reinstalled and the situation left in much the same state as it might have been had Burnes's advice been followed in the first place. The only positive gain was the annexation of Sind, which fell an inevitable casualty to the strategic demands of military operations so far beyond the British frontier. This advanced British rule across the Indus, but by now the painful lesson had been learned. The Great Game was not to be played in corps boots but in carpet slippers. Diplomatic pressures, exploratory feelers and, occasionally, short punitive forays took the place of territorial grabs and the heavy tramp of armies. Throughout the rest of the century Afghanistan was treated with great circumspection, heavily subsidised but never really trusted. The post to which Burnes had aspired, that of representative at Kabul, was never filled; the Afghans would not hear of it. And, so far as the Western Himalayas were concerned, no further attempt to explore them was ever mounted from beyond the Khyber pass. Afghanistan was as much closed to the traveller as Nepal or Tibet.

The effect of this was to give added significance to Wood's journey into the Pamirs. He was the first, and for nearly forty years the only Briton to penetrate them. No one was sent to verify his observations or to check on that puzzling confluence at Qala Panja. His work stood alone and, coupled with his own elusive career, constituted something of a geographical oddity. In 1873 when the British and Russian Foreign Offices got together to delimit their areas of influence in the Pamirs, it was to Wood's map that they turned. His was still the only authoritative account of the area and to this day the Wakhan corridor, that narrow finger of Afghan territory that separates Russia from Pakistan, follows the line of Wood's journey to the source of the Oxus.

The possibility that Sir-i-kol might not be the true source of the river was never quite lost sight of. Conscientious geographers and alarmist statesmen found this a disquieting thought which the conflicting reports of native surveyors, who visited the area in the 1860s and '70s, did nothing to allay. By the end of the century there were at least three claimants to be the parent stream of the Oxus. Depending on which you espoused, the demarcation based on the line of the main river could be understood to award to the British the whole of the Pamirs or none at all. To anyone with a tidy mind, a touch of Russophobia and a taste for geographical research, it was an inviting muddle.

George Nathaniel Curzon, Fellow of All Souls, MP for Southport, indefatigable traveller, brilliant scholar, remorseless protagonist and future Viceroy of India, was just such a man. He visited the Pamirs in 1894; his concern was not so much to explore the question as to exhaust it. Wood had ventured an opinion, Curzon now pronounced. He never actually saw Sir-i-kol or the confluence at Qala Panja, but he found a glacial source for the more southerly of the two feeders and, with irresistible logic and tireless reference to all the available evidence, declared it to be the true source.*

There the question rests. Curzon's *The Pamirs and the Source of the Oxus* is unquestionably a masterpiece of geographical research. The only doubt that assails the reader has nothing to do with his conclusions. It is just that one wonders whether the subject really justifies the labour and talent brought to bear on it. Matthew Arnold writes of the Oxus as having its 'high mountain cradle in Pamere', and most would be satisfied to know that it rises in the Pamir region from a number of sources, the most interesting of which is Wood's Lake Sir-i-kol though the most remote may indeed be Curzon's Wakh-jir glacier. Which of the many rills wriggling across that bleak terrain, sprawling and contracting, flooding and freezing with the seasons, should be dubbed the real Oxus seems a point of peculiarly doubtful significance.

The mistake was in picking on Wood's journey as the basis for a political demarcation. Rivers seldom make good frontiers and the Upper Oxus perhaps least of all. Had there been any follow-up to that pioneering journey of 1838 this would surely have been realised. The frontier, as elsewhere in the Himalayas, would have been based on watersheds, and there would have been no reason to call into question the claims of Sir-i-kol. With the passage of time and the indulgence of science, Wood's discovery would have become as unassailable as many another so-called source, hallowed by little more logical than tradition.

* The only paper on the sources of the Oxus which seems to have eluded Curzon was one by Gardiner. This is one of the old Colonel's classic productions. In mind-boggling detail he describes no less than nineteen feeders. As usual the challenge is to identify them and, as usual, there are just enough recognisable names and unassailable observations to suggest that it was not simply an elaborate fabrication. Gardiner did not try to prove that any of his feeders was the one true source, but one cannot but be grateful that Curzon never stumbled on the article. One shudders to think what a nasty mess the sharp stabs of his logic would have made of the Colonel's crazy ramblings.

PART THREE

The Karakorams and the Kun Lun
1841–1875

In the trans-Indus regions of Kashmir, sterile, rugged, cold and crowned with gigantic ice-clad peaks, there is a slippery track reaching northward into the depression of Chinese Turkestan which for all time has been a recognised route connecting India with High Asia. It is called the Karakoram route. Mile upon mile a white thread of road stretches across the stone-strewn plains, bordered by the bones of the innumerable victims to the long fatigue of a burdensome and ill-fed existence—the ghastly debris of former caravans. It is perhaps the ugliest track to call a trade route in the whole wide world. Not a tree, not a shrub, exists, not even the cold, dead beauty which a snow-sheet imparts to highland scenery.

T. H. HOLDICH. *Gates of India*

10. A Man of Science

Trekking in the Himalayas has been going on for longer than the tour operators would have one believe. Combined with a bit of casual shooting it was a popular holiday with the British officer and his family as early as the 1860s. Much earlier still, the first attempt at organising treks may be discerned in the activities of those three Aberdonians: Dr. James Gerard, who accompanied Burnes to Bukhara, and Captains Alexander and Patrick Gerard, who introduced Jacquemont to the mountains. In 1822, well before Burnes and Jacquemont and when Moorcroft was still in Leh, they were taking one of the first holiday parties up into 'the snowy range'. On May 7th they pitched camp at 7,000 feet on a spur of the Great Himalaya; 'Semla' they called the site after the name of a nearby village. It was a spot they knew well, and in the small hours they roused their party with the promise of an experience not to be missed. Bleary eyed and full of misgivings the trekkers followed them up a steep, well-wooded hill called 'Juckoo'. One of the party, Major Sir William Lloyd, left this memorable description of that morning.

We reached the summit of Juckoo long before daybreak and anxiously awaited the dawn. The sky appeared an enormous dome of the richest massy sapphire, overhanging the lofty pinnacles of the Himalaya, which were of indescribably deep hues, and strangely fantastic forms. At length five vast beaming shadows sprang upwards from five high peaks as though the giant day had grasped the mighty barrier to raise himself, while in the same instant the light rolled in dense dazzling volumes through the broad snowy valleys between them, and soon the glorious orb rose with blinding splendour over the Yoosoo pass, and assumed the appearance of a god-like eye. In a moment these rising solitudes flung off their nightly garments of the purest blue and stood arrayed in robes of glowing white. The intermediate mountains cast their disjointed dark broad shadows across the swelling ranges below, and the interminable plains were illumined, all the ineffable variety of earth became distinct; it was day and the voiceless soul of the great globe seemed to rejoice smiling.

This first description of the view from Simla is probably the finest. Dawn somehow intensifies the primeval desolation and the unassailable purity of the mountain wilderness. Lloyd looked on it with the wide-eyed wonder of one seeing the hitherto unseen and, without contrived art nor yet fear of the fanciful, he penned this grand evocation. It would be harder for subsequent visitors; these things look different through a crack in the chintz curtains.

Twenty-five years later when a certain Dr. Thomas Thomson was summoned from Ferozepur, he found on the ridges and spurs round Jakko Hill a sprightly and controversial little town. Since the days of Jacquemont's visit and the riotous rule of Captain Kennedy, Simla had come a long way. It was no longer the 'Capua of India' nor was it, as today, the Bournemouth. On the contrary, its hour of glory was nigh; it was about to become the summer capital of the British Raj. The first visit by a Governor-General was in 1827. Bentinck had repaired there often enough for a nearby peak to be christened 'Billy Bentinck's Nose' and, in the newly built Auckland House, Lord Auckland had finally reached his painful and disastrous decision to despatch the Army of the Indus. The Governor-General when Dr. Thomson arrived there in 1847 was Lord Hardinge. We have a description of him pacing the verandah of the same house as he read the morning's post.

Glancing up he might have seen that view of Lloyd's and thought no more of it than if it had been the wall of his study. Equally he might have seen that other celebrated Simla prospect, dense deadening mist which reveals nothing more noble than a dark conifer dripping silently into the dank grass. But, if he was lucky and happened to look up at just the right moment, his attention might suddenly have been arrested by the one glimpse which, even to people inured to the charms of that now prosaic panorama, manages still to convey something of the improbable situation of Simla. Without rhyme or reason the mist will suddenly compose itself and settle. It doesn't lift; it falls imperceptibly into the valleys. The outline of a nearby hill appears and is gone. Above it there is a brightness. It is as if the sun is about to break through. Keep looking and, high up but no longer far away, there shines down through a window in the mist, not the sun but a chunk of the Great Himalaya. Some nameless peak, some unclimbed crag or unmapped scarp it may be, but seen like this, the sun full upon its snowy hinterland and the bare black rock of pinnacle and precipice sharp against the sky and the busiest of men, even a Governor-General, must be roused.

A sobering thought to Lord Hardinge might have been the realisation that, though he ruled the destinies of a sub-continent, he must peer from his verandah on to a sea of mountains still largely unknown. The ridge of Simla is the watershed between the basins of the Ganges and the Indus. Half the town's drains empty themselves into the Giri river and thence into the Jumna, the Ganges and the Bay of Bengal. The other half, including those of Auckland House, descend precipitately to the Sutlej, the Indus and the Arabian Sea. Just as Kabul marks the westernmost extremity of the Indus basin, and so of the Western Himalayas, Simla marks the eastern limit. It was all very well to discourage hare-brained schemes of exploration from the distant security of Calcutta; quite different to deny a certain curiosity about what lay just beyond one's doorstep in Simla. Growing involvement in the affairs of the Punjab and Kashmir was shifting the fulcrum of British rule further to the north-west. This not only afforded some justification for the popularity of Simla but it also served to intensify speculation about what lay within and beyond the mountains. And although Afghanistan was now a closed door, there was still a chink of light towards Ladakh and Yarkand in the north.

Dr. Thomson found Simla in 1847 to consist of four hundred houses. Houses, be it understood, were British residences; the bazaar, where most of the native population lived, and the shacks at the end of the garden, where crowded together the servants and their multitudinous families, were not included. To build all the fanciful cottages and lodges with their dormers, balconies and rustic porches, the forests of cedar and evergreen oak had been whittled away. Already the town had that look of being spread-eagled along the bare crest of the ridge. It was no bustling city, but then neither is it today. The roads are too steep and narrow for much traffic, and the houses, secluded in extensive grounds, too far apart.

The view, the climate and the shooting all played their part in making the place a shrine for the British in India. It became a caricature of sylvan England, all rambling roses, crazy paving and casement windows, like a New Forest village, trees and all, wrapped over the contours of the Tyrol. Even Emily Eden, Lord Auckland's ever-scathing sister, grew to like it. 'If the Himalayas were only a continuation of Primrose Hill or Penge Common I should have no objection to spend the rest of my life on them.' But what most of all made the annual pilgrimage to Simla worthwhile was the society it offered. People went there because everyone else did. In the summer there were picnics, races, fairs and gymkhanas; in the winter, carols and Christmas parties with real snow outside and not some wretched sprig of mango but a proper spruce Christmas tree. They went, if they were honest, not just because everyone else did but because someone in particular did. The endless balls, dinners, at homes and evenings at the club provided a grand forum for the ambitious. Unashamedly they lobbied. Some liked to dance, to drink, to gamble, others just to watch and be watched; but all liked to lobby. This man sought a posting, that promotion, a third a wife and a fourth just an invitation.

They used Simla. And just as in its heyday the great attraction was the society so, now that that society is gone, the attraction of the place is hard to appreciate. Today, forlorn and tatty, it conjures up no images of splendour or excitement. Still there are the tea-rooms, the Gaiety Theatre and the clubs but it is hard to relish their associations. Kipling's gossipy matrons and heady young ladies were dismally parochial, and his raciest of rakes a pretty staid fellow. In the early days when Jacquemont slavered over the Perigord patés and Kennedy dispensed the dancing girls as freely as the champagne, the traveller could be excused for lingering awhile, charmed by the improbability of it all. But eccles cakes and awkward little flirtations in the saddle seem a poor substitute.

Dr. Thomson sounds as if he might have agreed. He was a rather serious-minded young Scotsman. For two and a half months he was detained at Simla but for all the notice he took of its inhabitants it might still have been virgin forest. In fact he would have preferred it like that, for his only interest was the flora. But then he had not come of his own free will; he had been summoned, and was now only concerned about pushing on as soon as possible. The delay was caused by the late arrival of his two companions. Together they were to constitute a border commission to visit Ladakh and delineate its eastern boundary with Tibet. The Governor-General had resolved to dispel some of the uncertainty about what lay beyond his verandah by taking advantage of an opportunity for the first official initiative in the Western Himalayas.

About this time, the mid-1840s, there was an apposite story going

13 Dr. Thomas Thomson

14 H. H. Godwin-Austen

15 T. G. Montgomerie

16 Robert Shaw

17 A Ladakhi village with Buddhist Chorten

18 In a Zaskari village near Padam

the rounds of the Simla bazaar. Like all good stories it had enough inherent possibilities to be true. There were also enough versions of it to keep one guessing where, if at all, the basis of truth lay. As told by a loyal sepoy in the Company's service it would have been Gulab Singh who, in early 1842, requested the meeting with Henry Lawrence. The Raja had pressing news and Lawrence, having himself some sobering reports for the Raja, readily agreed. Such a meeting was full of irony in the light of subsequent events. Gulab Singh, then Raja of Jammu and *de facto* ruler of Ladakh and Baltistan, was the greatest power behind the tottering throne of Lahore. But Lawrence, then Governor-General's agent in Ferozepur, was himself soon to assume administrative responsibility in Lahore.

In due course the meeting took place. Gulab Singh, with the un-hurried confidence of a man fanning himself with a royal flush, opened proceedings with a flood of extravagant compliments. Roses bloomed in the garden of perpetual friendship when the Governor-General's agent deigned to meet a humble ignorant Raja, and the power of Lahore was but as the moon beside the blazing sun of the British army. Lawrence, an impatient man, cut him short. What was the Raja's pressing news? With feigned ingenuousness Gulab Singh asked for the latest reports from Kabul. There had been none said Lawrence.

'Then it is news [said the Raja] sad news indeed. For there will be no more reports from Kabul. Burnes is dead, murdered by the Afghans. Sale's brigade is surrounded in Jalalabad. Kabul is fallen and Ghazni too. MacNaghten, Shah Shuja and the Army of the Indus—all are gone. The sun is in eclipse.'

Lawrence's thin face registered nothing. Friends of his would be amongst the captured and the dead. The British had sustained their worst ever defeat in Asia. But not a tear, not a prayer and certainly not a curse came from the imperturbable Lawrence. A man of the deepest courage and with an overriding religious sense of mission, he received the news in respectful and unflinching silence.

'If true, as I must reluctantly accept, this, Raja Sahib, is indeed sad terrible news. I respect your motives for calling me to hear it from your own lips and I thank you sincerely.'

He paused, apparently unable to say more. Then, as if pricked by a sudden and appropriate recollection, he stared hard at the Raja and continued:

'But will you too thank me for what I must now tell you? For I also am the bearer of sad terrible news. Out of Tibet by way of Nepal come the reports of the rout of another army. Six thousand strong it was. Now there are but a handful of survivors and they so frost-bitten as to be useless. The army was your army. Your general, Zorawar Singh, is himself among the dead. And the moon too is eclipsed.'

How Gulab Singh reacted to the news would depend on the narrator's powers of invention. But the meeting could well have some basis in history. The two disasters were almost simultaneous, and just as news from Kabul might be expected via the Punjab, so the defeat in Tibet first became known from the few survivors who straggled across the Great Himalaya into Nepal and the British dependencies between there and Simla.

Back in 1835, when Dr. Henderson visited Leh, Zorawar Singh—on behalf of his master, Gulab Singh—was poised for the conquest of Ladakh. By 1837, when Vigne tried to use Leh as his launching pad for the Karakorams, the Jammu Raja was in undisputed control of the country and in 1840, hard on Vigne's final departure, Zorawar Singh had overrun Baltistan. It was early the following year that he had invaded Tibet proper.

These conquests, though initially made in the name of Ranjit Singh, were not really Sikh achievements. In fact there is evidence that the Sikh governor in Kashmir actually tried to sabotage them. Gulab Singh himself was not a Sikh but a Dogra, that is a Hindu of the martial Rajput stock of Jammu. Even before Ranjit's death in 1839 it had seemed to Vigne that Lahore was rarely informed of his activities beyond the Pir Panjal and certainly after 1839 he was virtually independent. These should, therefore, be seen as Dogra conquests. Gulab Singh, no longer content to carve out of the Sikh empire a hill state for himself, was now engaged in the creation of his own trans-Himalayan empire. He still did not hold Kashmir but there was already talk of taking Lhasa and even Yarkand.

The invasion of Tibet in 1841 was so completely overshadowed by the Afghan disasters that it received scant attention. Yet as an epic of military adventure and the first ever Indian attempt to invade the Tibetan uplands, it is unique. It would have taken Alexander Gardiner, who was on the Dogra payroll at the time, to have done it justice, but sadly neither he nor any other foreigners took part in

Zorawar Singh's campaigns. Very little indeed is known about them. The attack on Tibet was probably prompted by a desire to plunder the Buddhist monasteries and to gain a monopoly of the shawl-wool production. This *pashm* was, as Moorcroft had pointed out, the most valuable commodity in the trade of the region. By a long standing arrangement most of it passed through Leh on its way from the high grazing lands of Tibet to the looms of Kashmir. But the Dogra conquest of Ladakh and the emigration south of many Kashmir weavers had diverted a small portion of the trade from Leh to Rampur, a British dependency just east of Simla. Zorawar Singh now hoped to close this outlet by adding to his control of the Ladakh carrying trade that of the Tibetan regions where the wool was actually produced.

With just six thousand men he set off from Leh up the Indus. Virtually no opposition was encountered and by the autumn of 1841 he held the whole of Western Tibet. The monasteries were systematically ransacked by Ghulam Khan, his Mohammedan lieutenant, who brought to the task all the iconoclastic fury of Islam. The flow of shawl wool to Rampur dwindled to a trickle, and Dogra rule reached to the Nepalese frontier and included the centres of Rudok, Gartok and the sacred lake of Manasarowar.

Whether Gulab Singh himself was responsible for the invasion, or whether it was simply the idea of the over-zealous Zorawar, is not clear. But the implications were far from being a domestic Dogra affair. The Chinese, whose grasp on Tibet was then rather firmer than it became at the end of the century, were unlikely to distinguish between the activities of an Indian Raja and those of the British rulers of India. The British government was then in the process of negotiating a peace with China. It was a delicate state of affairs which the Dogra moves in Tibet would surely jeopardise, Added to this there was a question mark over the attitude of the Nepalese who, whether they befriended their new Dogra neighbours or with Chinese encouragement attacked them, would greatly embarrass the British policy of keeping Nepal isolated.

Thus news of the Dogra advance elicited a strong protest from the British government. Gulab Singh replied with a cheeky proposal for a joint Anglo-Dogra attack on Western China. This was rejected and reluctantly, but with an eye to the advantages of British friendship, he agreed to withdraw from Tibet. A British officer, Lieutenant Joseph Cunningham, was despatched up the Sutlej to observe and report on the disengagement. He reached Rampur in the late autumn but failed to penetrate Tibet in time for the bloody conclusion of the campaign.

With slow but impressive resolve the Chinese had finally reacted. An army of about ten thousand left Lhasa in October. In November they routed the small detachments which Zorawar Singh contemptuously sent to oppose them, and in early December they were within sight of the main Dogra force. Alexander Cunningham, the brother of Joseph, pieced together the best account of what happened.

> The two armies first met on the 10th of December, and began a desultory fire at each other, which continued for three days. On the 12th Zorawar Singh was struck in the shoulder by a ball, and as he fell from his horse the Chinese made a rush, and he was surrounded and slain. His troops were soon thrown into disorder, and fled on all sides, and his reserve of 600 men gave themselves up as prisoners. All the principal officers were captured and out of the whole army, amounting with its camp-followers to 6,000 men, not more than 1,000 escaped alive, and of these some 700 were prisoners of war.

> The Indian soldiers of Zorawar Singh fought under very great disadvantages. The battlefield was upwards of 15,000 feet above the sea, and the time mid-winter, when even the day temperature never rises above freezing point, and the intense cold of night can only be borne by people well covered by sheepskins and surrounded by fires. For several nights the Indian troops had been exposed to all the bitterness of the climate. Many had lost the use of their fingers and toes; and all were more or less frostbitten. The only fuel procurable was the Tibetan furze, which yields much more smoke than fire; and the more reckless soldiers had actually burnt the stocks of their muskets to obtain a little temporary warmth. On the last fatal day not one half of the men could handle their arms; and when a few fled the rush became general. But death was waiting for them all; and the Chinese gave up the pursuit to secure their prisoners and plunder the dead, knowing full well that the unrelenting frost would spare no one.

The prisoners included Vigne's old friend, Raja Ahmed Shah of Baltistan. For two years he had been held by the Dogras and now duly rejoiced at their defeat. The Chinese treated him well but, when it became clear that there was no hope of being restored to his throne, he fell to pining for his lost home in Skardu and died soon afterwards. Also captured was Ghulam Khan, the desecrator of the Buddhist monasteries. He fared less well. He was tortured with hot irons, his flesh was torn off in small pieces with pincers and the mangled remains still quivering with pain were left to experience whatever residual agonies a protracted death might offer.

The Chinese forces pressed on into Ladakh and, for a short time in early 1842, actually held Leh. Then the Dogras returned in force, and a treaty was signed between the two parties restoring the traditional boundaries and inferring that the Leh monopoly of the shawl wool trade should continue. If this trade was the chief motive for the invasion then the Dogras certainly won the peace, if not the war. From a British point of view the situation, though not as fraught as in 1841, was scarcely satisfactory. There was no telling when Gulab Singh might again interfere in Tibet and the British attempt to attract some of the wool trade had been foiled. But, as soon as affairs in Afghanistan were settled, relations with the Punjab worsened and it was not till the end of the first of the two Sikh Wars in 1846 that a further opportunity to intervene arose.

During these momentous events of the early 1840s Gulab Singh can be seen playing his cards with the utmost skill. He lacked the magnetism of his old master, Ranjit Singh; his charm inspired terror rather then affection. But there is no questioning his ability. His eye was ever for the main chance and the rule he founded, though less notable, was to last considerably longer than that of the dead Maharaja. Whether or not he actually passed on the sad news of the disasters in Afghanistan, it was to the British that he increasingly turned. He had listened to their advice about withdrawing from Tibet, and in 1842 he co-operated in the reprisal invasion of Afghanistan. In 1845 he significantly failed to assist the Sikhs, still nominally his overlords, and thus facilitated the success of the British forces. By way of reward he at last gained control of the Kashmir valley and was recognised by the British as Maharaja of Kashmir, Jammu and Ladakh.*

* It is not apparently correct to say that the British sold Kashmir to Gulab Singh. Nor to say that the British never ruled Kashmir. They did, for a week. By the Treaty of Lahore, Kashmir, Jammu and Ladakh were ceded to the British by the defeated Sikhs in lieu of an indemnity for the war. By the Treaty of Amritsar a week later Gulab Singh was rewarded by being recognised as Maharaja of these territories. Several important conditions were imposed. Payment of the indemnity was one of them, but was of little significance compared with the political and strategic implications of the others.

The treaty of Amritsar, which confirmed this arrangement, reserved to the British certain rights which have an important bearing on the penetration of the Western Himalayas. In the first place, though recognised as an independent sovereign, the new Maharaja was bound to accept British arbitration in any dispute with his neighbours. In return he was promised British assistance in defending his territories. This could be seen as reserving to the British a say in relations between Ladakh and Tibet. Equally it could be seen as another forward move in the Great Game. Kashmir remained nominally independent, but the frontier which the British were now committed to defend had thus been pushed hundreds of miles further into the mountains.

Just exactly where it now lay was far from clear. The Treaty of Amritsar was wonderfully imprecise. 'All the hilly or mountainous country, with its dependencies, situated to the eastwards of the river Indus and the westward of the river Ravi' was ceded to the Maharaja. This oft-quoted description betrays an ignorance, inexcusable by the 1840s, of the general lie of the land. Through the mountains both rivers pursue a predominantly east–west course so that scarcely any place can be said to be east of one and west of the other. Areas specifically mentioned in the treaty, like Ladakh, are actually bisected by the Indus. And the only regions clearly bounded east and west by these two rivers are not mountainous or hilly but a dead flat wedge of the plains.

Something would have to be done to define the limits of the new state more accurately and, fortunately, provision for this was also made in the treaty. The precise extent of the Kashmir state was investigated and discussed for the next fifty years and played no small part in the exploration of the region. Moreover the vague wording of the treaty was found to have its advantages. Once 'eastward of the river Indus and westward of the river Ravi' was seen as nonsense, 'the hilly or mountainous country with its dependencies' could be taken to include as much of the mountain complex, stretching over the Karakorams to the far Pamirs and Hindu Kush, as was convenient.

More immediately, the corner of the new state which bordered the British dependencies and Tibet needed to be sorted out. Here a continued desire to tap the *pashm* trade had prompted the British to retain a corner of the Sikh empire, namely the small states of Kulu, Lahul and Spiti. In 1846 Alexander Cunningham explored the watershed between these new acquisitions and Ladakh, and laid down a frontier based on it.

The following year Cunningham was again deputed to Ladakh. Dr. Thomas Thomson, waiting in Simla, was to be one of his companions and the other, Lieutenant Henry Strachey. Ostensibly the object was to define the boundary between Ladakh and Tibet so that Gulab Singh might never again have the excuse of a frontier dispute for re-invading Tibet. But there were also well-founded rumours for regarding the 1847 Boundary Commission in quite another light. The Royal Geographical Society got wind of these and welcomed the project as 'an exploring expedition'. Their information was that the party was heading for Yarkand and intending to spend the winter there or in Kotan. Furthermore the members of the expedition were then by different routes to proceed east, to rendezvous in Lhasa and to regain India by following the Tsangpo till it swept round the Himalayas and emerged in Assam as the Brahmaputra. If the Society was right then this most ambitious expedition represented the first official attempt for fifty years to penetrate Tibet and the first ever to cross the Western Himalayas.

A year later the President of the Society changed his tune. He referred to the expedition as a Boundary Commission and said no more about Lhasa and Yarkand. But, in fact, the original plan had been every bit as ambitious as he had imagined. No official expedition ever had a freer hand than did Cunningham, Strachey and Thomson. Reading the instructions issued by the Governor-General it is almost as if the boundary work was to be simply a front for uninhibited exploration.

There were two stipulations. They were not to be away for more than two years, and they were on no account to cross the 'Bolor Tagh' mountains. Since this might lead to 'collision with the bigotted and jealous Mahometans of Independent [i.e. Western] Turkestan', the Bolor Tagh must mean the Pamirs. Otherwise it was up to them. Once the boundary work was completed, and the Governor-General admitted that with so little in dispute it should not take long, they were 'individually to use their best endeavours to increase the bounds of our geographical knowledge'.

Winter, it was suggested, might profitably be spent beyond the mountains in Yarkand or Kotan or, failing these, in Rudok 'on the borders of the Great Desert'. Come spring Cunningham should return to the Indus and follow its course down to Gilgit, and then continue west through the so-called Dard* countries of Yasin and Chitral. Strachey might concentrate on Tibet, heading 'as far eastwards as he can go, even to Lhasa', and return via Darjeeling or Bhutan. Only Thomson was given a prosaic task. He was to 'employ himself ascertaining the mineral resources along and within the British frontier'. He did rather more than that, but for the present it is sufficient to note that the British frontier was taken to mean not Kulu and Lahul but the supposed Kashmir frontier on the Karakorams.

With a brief like this imagine what Moorcroft or Burnes might have achieved. If the Chinese had proved obstinate, Burnes would have slipped past dressed as a half-witted lama. Moorcroft, on the other hand, would have put it about that he was the Queen's own emissary, and have swept on with flags flying. Neither the exceptionally severe winter of 1847–48 nor the outbreak of war in Gilgit would have stopped them. Here, after all, was the long-awaited official initiative for prising open the back door into China, for establishing a British interest in the existing Himalayan trade, for gaining the sort of geographical information about the lands beyond the mountains that had been found so vital in the Afghan debacle, and for bringing these places within the scope of official British policy. It appeared to be a legitimate opening and one not to be fluffed by an over-nice respect for Chinese sensibilities.

Cunningham and Strachey joined Thomson in Simla during the summer of 1847, and the Commission finally got under way on August 2nd. It was the middle of the monsoon when the mist is so thick you can hardly see across the Mall—no time for a grand send-off. Half the expected porters failed to materialise, and they were thankful that their disorganised departure went unnoticed. Their route lay up the Sutlej valley to Spiti and thence to Ladakh. It was much too late for a full season of demarcating the boundary but, if the Chinese and Dogras were ready and if the job proved as simple as Hardinge had anticipated, they might yet be able to seek winter quarters further afield.

The idea that they would be allowed into Yarkand or Lhasa was based on the assumption that the Chinese would welcome the demarcation of their Tibetan frontier. This was a serious miscalculation.

---

* Although the word Dard was never acknowledged by any of the peoples of the region, it was identified with Ptolemy's Deradrae and widely used during the nineteenth century to describe the peoples living west of the Indus, especially in Gilgit, Chitral, Yasin and Chilas.

The Chinese had few anxieties about Gulab Singh and never welcomed attempts to circumscribe their sphere of influence; the nebulous character of Chinese and Tibetan suzerainty scarcely lends itself to the clear-cut European notion of territorial integrity. Nor was the other inducement offered to the Chinese, the commercial advantage of diverting the shawl-wool trade through the new tariff-free corridor of British territory, particularly enticing. The trading relations of Tibet were based on tradition rather than profitability and they followed closely ancient lines of religious and political intercourse. Overtures from the barbarian newcomers in India were not likely to be allowed to prejudice the traditional connection with Buddhist Ladakh. Added to all this there was the difficulty of approaching the Lhasan authorities. The invitation to them to send their own boundary commissioners went via Gartok but whether it was actually delivered there, whether it was couched in intelligible Tibetan and whether it was ever forwarded to Lhasa, are all doubtful. Another approach was made direct to the emperor via Hong Kong, but this too probably never got further than Pekin.

However, the surprising news was that two officials had already reached Gartok and, buoyed up by this, the Commission made good speed to Spiti. There, in an attempt to take a short cut across Tibetan territory, they received their first check. The Tibetans were waiting for them and, as usual, refused entry. Thomson noted, 'I have no doubt that if we had resolutely advanced no serious opposition to our progress would have been made . . .' Trebeck and Jacquemont had both called the Tibetan bluff and had got away with it. A resolute move at this stage might have made all the difference. It would not have conjured up a party of friendly Chinese commissioners, but it would have shown they meant business and perhaps opened up direct communications with Gartok. But Cunningham was a cautious man. He had been warned by Lawrence, now Resident at Lahore, to avoid any risk of a collision and, though Hardinge had had the last word, telling him in effect to ignore this over-cautious advice, he chose to stand down. The Commission meekly withdrew and reached Hanle in Ladakh by a roundabout route over the lofty passes of Lanak and Parang.

True, they were still hoping for Chinese co-operation, but at this encounter they also learnt that the Tibetans knew nothing of any Chinese border commissioners. At Hanle this news was confirmed. The Chinese officials who had supposedly arrived at Gartok now turned out to be a solitary Tibetan revenue officer who had already set off back to Lhasa. Even the Dogra commissioners had not yet appeared. Cunningham took this as the signal to abandon the boundary question altogether and to concentrate on the more interesting part of their instructions, to 'individually use their best endeavours to increase the bounds of our geographical knowledge'.

On the march

The Chinese being so uncooperative there seemed little hope of reaching Yarkand or Rudok but, with winter pressing on—it was already mid-September—they each now went their separate ways. Strachey turned east. His interests were purely geographical and they centred on Tibet. For the next ten months he prowled along the Ladakh–Tibet frontier. His exact itinerary is lost, but he appears eventually to have crossed the south-west corner of Tibet and returned to India near Moorcroft's Niti pass. He thus connected his observations in Ladakh with those he and his brother had previously made near Lake Manasarowar. The results of his travels were embodied in a weighty and authoritative official report entitled *The Physical Geography of Western Tibet*. This was later published and for it he was awarded one of the Royal Geographical Society's Gold Medals in 1852. The strength of this almost unreadable document lay in its attempt to relate the rather poetic ideas of traditional Tibetan geography to the physical configuration of the country. Western Tibet he therefore took to include not just the Rudok and Gartok regions but Ladakh and Baltistan as well. The result, in so far as it inevitably tends to emphasise the integrity of the whole region, is a curious memorial to one who was supposed to be defining a boundary across it.

Cunningham and Thomson travelled on together as far as Leh. Thence they intended to proceed by different routes to Gilgit. Thomson was to investigate the mountains to the north, while Cunningham took an easier more southerly route skirting Kashmir. If his companions had rather narrow interests Cunningham's wide range of enquiry more than compensated. When the winter turned out to be exceptionally severe, he welcomed it as an excuse to abandon the idea of Gilgit, and to concentrate on the antiquities of Kashmir. They had been told to pursue their own interests; Cunningham took this literally. He arrived back in Simla in 1848 with a camel load of Buddhist statuary, three unknown Sanscrit dramas and 'the oldest dated inscription hitherto found in India'. It was not quite what the Governor-General had expected from the leader of his Boundary Commission. However, Cunningham made amends by producing a report on Ladakh which remains to this day the standard work on the country. The format is the usual one for an official gazette, a chapter for geography, another for history, another for communications, climate, productions—animal, vegetable and mineral—and so on. What distinguished it was the degree of scholarship, embracing anthropology, language, ritual, literature and archaeology. Taken in conjunction with Strachey's work, Ladakh could be said to be not just explored but exhausted. The 1847 Boundary Commission had failed dismally to define the boundary, or, according to the official summary, 'to accomplish any of the political purposes for which it was appointed'. But they had succeeded brilliantly within the narrow limits finally accepted.

But what of Dr. Thomson? Not much was to be expected of him as an explorer. He was there as a distinguished naturalist and the party's surgeon. If there was no opening for him like that which took Lord to Kunduz, he could always be left to potter along the frontier rooting for rocks and plants. The son of another eminent scientist, he had discovered fossilised molluscs in the Clyde at seventeen and pectic acid, the vital gelatinous constituent of fruit jellies, in carrots at nineteen. Arrived in India, he had embarked on a study of the flora of Afghanistan and only abandoned it when, as a prisoner after the siege of Ghazni, he was about to be dispatched to the Bukhara slave market. He was indeed a dedicated man of science. Nevertheless he resented being written off as an explorer. Though alarmingly susceptible to the effects of altitude and totally inexperienced as a surveyor, he willingly took on the most arduous assignment of the whole expedition, that of exploring the northern confines of Ladakh towards Yarkand and Gilgit. Where Moorcroft, Wolff and Vigne had all failed, the good doctor was to try again. Some idea of his inexperience can be gained from Cunningham having to lend him a sextant and compass and then show him how to use them.

He left Leh on October 11th, aiming to cross the Karakoram pass and advance a 'few marches on the northward towards Yarkand'. But, just like Vigne before him, he reached Nubra much too late in the year for 'exploration at great altitudes'. Turning back, he followed the Shyok and Indus down to Skardu with the idea of continuing on down the Indus to Gilgit for his expected meeting with Cunningham. Again he was thwarted. Gilgit was 'not in a fit state for scientific investigation'. There were rumours of trouble between the Dogras and the natives. On December 2nd he turned back towards Kashmir for the winter. Vigne's route over the Deosai and Burzil passes was already closed and his only exit lay back up the Indus and round via Dras and the Zoji La. He spent December 4th, his thirtieth birthday, edging along the cliffs below Tolti, the twelfth fighting through blizzards and snowdrifts into Dras and on the fourteenth, for the third time, was forced to retrace his steps. The pass was closed and he was trapped. He gave up his tent in favour of the native cowhouses, and headed back to Skardu, arriving there on Christmas Day.

He was the first European to winter in Baltistan. He kept a careful check on the weather and, while avalanches boomed all round, sorted out his collection of wild flowers and rocks. Strachey managed to send down to him from Leh some brick tea, 'not superexcellent in quality', and some sugar. Otherwise he relied on local supplies. At the end of February he made a second try for Gilgit, reaching the dread gorges below Rondu before firm reports of open warfare between Gulab Singh and the Gilgitis dissuaded him from going further. He retired again to Skardu, continued up river to Dras and finally crossed the Zoji La into Kashmir and then the Bannihal pass to Jammu.

By now one would have thought he had had enough. For nine months he had been travelling almost continually and for the last six, during the coldest of Himalayan winters, he had been entirely on his own. He was not a lover of solitude or the wide open spaces, nor was he of a particularly rugged constitution. His portrait shows a pallid man of distinctly delicate habits who would not last long without a bath and clean linen. Yet he had already requested permission of Lord Dalhousie, the new Governor-General, to continue his work in the mountains. He wanted to try a new route over the Pir Panjal and Great Himalaya to Ladakh and, what was more, he wanted to have another go at getting beyond Nubra on that daunting track to Yarkand.

Dalhousie, no doubt hoping that somewhere the Commission might yet break new ground, agreed; for the fifth time Thomson turned back on his tracks. It was May 23rd 1848. The season was young enough, the ground well enough prepared and Thomson now experienced enough for what Sven Hedin calls 'one of the most important and successful [journeys] ever undertaken against the secrets of the highest mountainland on the earth'.

The route he had chosen to pioneer into Ladakh was that via Chamba and Zaskar. It seemed as direct as Moorcroft's Kulu-Lahul route and was said to be much used by native travellers. To escape the heat in the plains, he tacked through the outer hills to Chamba and then climbed the 15,000-foot Sach pass over the Pir Panjal. The Sach pass, and the Pangi region beyond, is still mule country. These days buses trundle over the Rohtang, Bannihal and Zoji but here there are no roads, no jeeps. The traffic sounds are the resonant ding-ing of mule bells and the fussy tinkling of sheep bells.

For the botanist there is no better place than the Pir Panjal and no better time than early summer. The pass is approached through deep woods of oak, sometimes festooned with vines. It was summer when Thomson entered them but emerging from their shade on to the bright grassy slopes above was like stepping back into the spring. Rhododendrons are in full flower in June, especially the little yellow *campanulatum*, the pride, in Thomson's view, of the Indian mountains. The grass is dotted and finally smothered with primulas and potentillas. Once on the snow there is still a wealth of interest, for wherever a rock has thawed its way through, there, right to the edge of the soft snow, springs up the same bright carpet of Alpine flora.

In the doctor's narrative there are no bold strokes, no purple passages. Plant by plant, rock by rock, he painstakingly constructs his description. To appreciate its effect and to gauge the depths of his enthusiasm, one must stand well back. One must measure the rich and sensitive treatment of this stage of his journey against his narrative as a whole. To the naturalist it was definitely the highlight of the trip.

From the Sach pass Thomson descended to the Chandrabhaga (Chenab), and followed that river west through Pangi as far as the Gulabgarh, and then turned north towards Zaskar. The grand scenery of steep pine-clad slopes, sharp mountain profiles and well wooded valleys gave way to the higher, dryer and barer terrain of Ladakh. It is surprising how much of interest a botanist can find in such a country though, with the bones of the earth comparatively bare of vegetation, even of soil, it is the geologist who now comes to the fore. Thomson's narrative dawdles through the moraines and alluvia. He cracks open conglomerates and sandstones, puzzles over schists, slates and limestones, and pockets bits of mica, gneiss and basalt. Higher still, as he climbed the Umasi La over the Great Himalaya, the ground beneath his feet became beds of snow and glaciers contoured with fissures.

What from a distance looks like a postcard snowscape becomes something more varied and intricate to anyone crossing it. It is never just snow, but new snow or old, hard or soft, resting on ice, a glacier or more snow. It may be steeper, deeper, more crevassed and, commonly, more treacherous, than it looks. On the other hand it may be a welcome relief from the boulder-hopping of a moraine, or the slithering traverse of a loose shale slope. There is no easier going than along a smooth gently sloping glacier or over the crust of a well frozen snowfield. Thomson was becoming quite a connoisseur of these things. It provided a new field for scientific observation when rocks as well as plants were buried out of sight.

As the swirling snowflakes reduced visibility to ten yards, he crouched on the crest of the Umasi La and got his water to boil at 180 degrees. This meant an altitude of 18,000 feet. Ahead lay Zaskar, the southernmost district of Ladakh. The journey across it from the Umasi La to Leh took nearly four weeks. This was far longer than he had expected, and was entirely due to the rugged nature of the country. In Zaskar it is seldom possible to follow the rivers which drain north towards the Indus. The ground is too cut up, and the track forever deviating to scale yet another lofty pass well away from the desired line of march.

Thomson was not the first to emphasise the parallelism of all the Himalayan ranges, but he wrote of this feature with understandable feeling. Zaskar combines all the difficulties of climate and altitude common to Ladakh as a whole with the precipitous character of the mountains on its perimeter. There are no open plains as along the Indus or further east. All is mountainous and since the mountains, in so far as they have any system, trend parallel to the Great Himalaya, and since Thomson's route lay from south to north across them, he was actually tackling the stiffest of all possible routes to Leh. Occasionally he ground to a standstill and held up his hands in horror at the relentless switchback.

> I find it extremely hard to describe in an adequate manner the extreme desolation . . . The prospect before me was certainly most wonderful. I had nowhere before seen a country so utterly waste . . . Directly in front across the Zanskar river a rocky precipice, worn and furrowed in every direction and broken into sharp pinnacles, rose to the height of at least two thousand feet, overhanging a steep ravine, while to the right and left mountain was heaped upon mountain in inextricable confusion, large patches of snow covering the higher parts.

In Leh he found Strachey just back from another visit to the Tibetan frontier. He rested for a week and on July 19th started on the last stage of his journey. His destination he gave as 'the mountains north of Nubra'. Beside the evidence of a modern map this sounds pretty vague, but one must remember that Thomson, and Cunningham and Strachey, had no idea of the depth of the mountains that still intervened between Leh and Yarkand. The only known accounts of the route were the brief notices submitted by Moorcroft's agent, Mir Izzet Ullah, back in 1813, and by Burnes's informant in 1832. From them Thomson knew of the Karakoram pass and knew that it was on the watershed between the Indus basin and the rivers of Turkestan. The range it crossed was assumed to be a continuation of the mountains north of Baltistan. Strachey, with cautious circumlocution, gave them no name at all, while Cunningham went to the other extreme, naming 'what we call the Bolor and Karakoram . . . which probably merges into the Kuen Luen in the east', the Trans-Tibetan chain. Thomson noticed that the range further west was called Mustagh, above Nubra it was called Karakoram and might conveniently be referred to, as a whole, as the Kun Lun. In other words he, like the others, knew of only one mountain system. That the Karakoram pass had nothing whatever to do with the Kun Lun, or that the mountains to the south of Yarkand known by that name were yet another vast separate system, is not even hinted at.

'The mountains north of Nubra' were, therefore, thought to be the same as those south of Yarkand. Thomson imagined that by scaling the Karakoram pass he would be overcoming the last barrier between India and Central Asia, and he hastened on towards it. He crossed the Ladakh range between Leh and the Shyok by the 17,500-foot pass of Khardung, the first of the five classic passes on the route to Yarkand. Not long before it had been possible to by-pass at least one of these by following the S bends of the Shyok. Now that route was blocked by glaciers which had nosed right down into the valley and pushed across the river bed into the precipices on the other side. Besides which, the river in July was in spate. It would have been impossible to follow up its bed and even crossing it proved extremely difficult. Though seldom able to ride, Thomson still had his horse. Crossing the Shyok, it took four men to support and guide the beast against the racing current.

To rejoin the river in the middle of the S, he proceeded up the Nubra valley and climbed towards the second of the great passes. This was the notorious Sasser, not the highest but probably the most impressive and dangerous. In Nubra, which by Ladakhi standards is a lush and populous spot, he stocked up with twenty days' rations for the uninhabited regions ahead. The ascent was 'exceedingly steep, almost precipitous'. The track zig-zagged back and forth and was so littered with bones and skeletons as to appear from a distance to be paved white. Mostly they were the remains of mules that had died on the journey, but there were human skulls as well. Later travellers reckoned that, providing there was not a fresh fall of snow, guides were a waste of money along this macabre trail. You simply followed the bones.

Thomson was now on ground untrod even by Moorcroft or Vigne. From the top of the first ridge he gazed towards the Sasser. '. . . I was able to see something of the road before me regarding which I had previously had little information except in accounts of its extreme difficulty. These I had inclined to consider exaggerated, but the prospect before me now was undoubtedly far from tempting.' The Sasser turned out to be a network of vast glaciers. Some he reckoned were at least five miles long and between them lay moraines several hundred feet high, composed of boulders as big as houses. One good thing about Thomson's narrative is that it is never open to doubt. When he talks of desolation he means it; the day goes by without sight of a single plant or blade of grass. The savants had their doubts about Vigne's descriptions of the Karakoram glaciers. When Thomson's experience of the Sasser confirmed them, there was an end to the matter.

The altitude of the pass was about 18,000 feet. He didn't measure it but made a rough calculation and the return of a severe headache confirmed it. On the descent it started to snow; he was lucky to have got across before the weather changed. Even in July whole caravans have perished on the Sasser in a squall. Back down to the Shyok, across it and up yet again to meet it at the top of the S. There was no pass to cross this time. After a long but gentle ascent he found instead an 'open, grassy, somewhat undulating plain'. It was a good five or six miles across and its lowest point was about 17,000 feet. A remarkable phenomenon, thought Thomson; he wondered whether it was not the highest plateau in the world. In fact it probably is. It is called the Depsang plateau and, like the Pamirs only more so, it always struck subsequent visitors as 'the veritable top of the world'. Dr. Bellew, who crossed it on the way back from Yarkand in 1874, left the following description.

All around appeared mountain ranges, none of which were less than 20,000 feet high, whilst to the west rose two peaks of much greater height; yet in the distance they seemed below us, for the land around sloped away down on all sides. In whichever direction we looked the sky appeared below us and the world slunk out of sight. In fact we felt as if we had risen above the world and were now descending to it in front of us. The Karakoram left behind us appeared like a mere crest on the undulating surface of the country and the mountain ranges in front and on all sides seemed to struggle up from below to reach our level.

Nothing so fanciful is to be expected of Thomson who was more exercised by the variety of pebbles. He was also suffering severely from the altitude with a continuous headache which grew unbearable with the slightest exertion. The next geologist to take this route, a companion of Bellew's, actually died on the Depsang from the effects of altitude. The Shyok was crossed for the last time, and on August 18th he left his horse tethered and his tent standing and set off for the crest of the Karakoram pass.

Originally he may have intended to cross it. He doesn't say much in his book, but in a letter to his brother he wrote of the danger of arrest at the first Chinese post and the probable difficulty of obtaining supplies on the other side. He also made enquiries about a possible route to Kotan, which might take him round any Chinese posts. The twenty days' food he had taken would certainly have been sufficient to see him across even the Kun Lun. It looks then as if the real reason for calling it a day was his own health. He repeatedly mentions the severe headaches, added to which altitude has a general debilitating and demoralising effect.

Just possibly he might still have gone on if what he calls 'the prospect before me' had been tempting enough. Foolish as it is to expect some new and shattering vista from the crest of every pass, the traveller always does. Scarcely daring to look up, he concentrates rock by rock on the progress of his feet. He seems to move with the absurdly slow deliberation of an astronaut. Breathing is agony, the head pounds and the calf muscles scream. The top is the reward, always exciting because it has to be. Every other reason for going on seems irrelevant, and only an immediate objective can justify the appalling exertion.

Thomson had the added incentive of knowing that this was not just any pass. He was about to be the first European to stand on the Karakoram watershed between India and China. To the best of his knowledge the ground beyond sloped gently down to the fabled cities of Yarkand and Kotan. From some of the passes over the Pir Panjal you could see the minarets of Lahore. Might there not be a comparable view to the north from the crest of the Karakoram? Might he not spy some distant speck that was one of the walled cities unseen by Europeans since the days of Marco Polo?

The ascent, though agonising, was not steep. Quite suddenly he found himself on the top. It was just a rounded ridge, swept free of snow by the wind and bounded by steep slopes on each side. No grand gateway, this, but a grim and forbidding cleft. The height he made 18,200 feet, but still it was no vantage point. 'Towards the north, much to my disappointment, there was no distant view.' He could follow the track down for just half a mile, then steep mountains higher than the pass closed the view. To a sick man it was not inviting. Nor was there any consolation for the naturalist. Not even a bold little saxifrage crowned the pass. The only signs of life seen all day were 'ravens, a bird about the size of a sparrow, a bright metallic coloured carrion fly and a small dusky butterfly'.

Thomson however was content. He had achieved more than was ever expected of him. 'I think I have determined', he told his brother, 'the points of most interest both geographically and botanically.' Spurning generalisations, exaggeration, hearsay and imaginative description, he stuck to what he saw and what he recognised. The scientific habit makes him a dull writer but a great pioneer. His observations, particularly on the part played by glaciers and lakes in moulding the structure of the mountain region, and on botany and geology in assisting the geographical interpretation of it, were accepted without question.

'The remainder', he continued, 'will be done some day from Yarkand but cannot till the Russians take it from the Chinese.' Even he was mindful of the power struggle for Central Asia, though the thought that his journey had any political significance seems scarcely to have crossed his mind at the time. There were, however, those who immediately derived some satisfaction from hearing that the Karakoram pass did not seem to be the end of the mountain barrier to the north. The ramparts of the new frontier were indeed formidable. When, in 1866, the Royal Geographical Society belatedly decided to give Thomson a Gold Medal it was as much in recognition of the increasing political significance of his journey as of its scientific achievements. But by then Thomson was ensconced in London's Kew Gardens rightly enjoying the reputation of one of the greatest botanists of his time. Others had taken up the challenge of the Karakorams and the Kun Lun.

11. Plane-Tables from the Hills

The Karakoram pass scaled by Thomson was first actually crossed by two Germans, Herman and Robert Schlagintweit, in 1856. The credit could, of course, go to Gardiner but, without quibbling over his nationality or casting any more slurs on the authenticity of his travels, the fact is that his account is so meagre that even his route is in doubt. Not so with the Schlagintweits. They followed in Thomson's tracks precisely, crossed the pass and then veered off east towards Kotan. In the history of Himalayan exploration they deserve the fullest treatment. They went on to cross the Kun Lun, the first to break through the entire mountain barrier between India and Turkestan, and the first to distinguish the Karakorams and Kun Lun as separate mountain systems. In the following year a third brother, Adolph, repeated the feat and actually reached Yarkand and Kashgar. But he never returned. He blundered into a civil war and was assassinated just outside Kashgar.

Tremendous achievements they were, yet these journeys received scant attention at the time. So much so that the next Europeans to reach Yarkand liked to imagine that they themselves were the first. It would be a pleasure to redress the situation; but sadly this is not possible. One reason is that this is a story of explorers, not a history of exploration. The most self-effacing travellers, like Wood or Thomson, may stand out clearly, revealed rather than dwarfed by their exploits. They are men, 'flesh and blood and apprehensive'. The Schlagintweits are more like some impersonal machine; the individuals are quite crushed by their achievement. The mighty volumes which record their observations reveal everything about their work. We know a very little about their careers but we learn absolutely nothing about their personalities. Gardiner could be sparing of the facts but the man himself is unmistakable. It is the other way round with the Schlagintweits. They took on the whole of India, aiming to reduce it to a gigantic table of observations. Of hot springs alone, for example, they listed and described 656. In the process they damned themselves to a regrettable obscurity.

In all honesty another reason must also be conceded. The story of their travels has never been translated into English. The six quarto tomes of truly Himalayan proportions, their pages of heavy gothic script still uncut, may conceal those pearls the biographer cherishes. Maybe there is somewhere in that deep and turgid narrative a basis for characterisation. Nearly drowning in the process, the present author has searched and searched in vain.

Back in the 1850s they had sounder reasons for ignoring the work of the Schlagintweits. For a start the servants of the Honourable Company had other things on their minds. Returning from the Karakoram pass in 1848, Dr. Thomson found himself cut off again, this time in Kashmir; the Second Sikh War had begun. It lasted only till the following year, but the annexation of the Punjab that resulted was a colossal undertaking which was carried out with single-minded zeal. It was promptly followed by the Central Provinces (now Madhya Pradesh) and Oudh (in Uttar Pradesh) also coming under direct British control; the last great wave of British expansion in India was breaking in grand style. Here was more than enough to absorb the attention and tax the energies of the restless or the inquisitive. Then, when the brothers had completed their travels and when Adolph's murder might have won them widespread attention, it was 1857 and the beginning of the Indian Mutiny. The execution of a lone German in Kashgar could scarcely compete with the wholesale massacre of men, women and children that was sweeping the British cantonments from Bihar to Lahore. In this, the most traumatic event in 200 years of British rule in India, an almost inevitable casualty was the Honourable Company itself. Over the years it had increasingly shared its responsibilities with the government in London, but now in 1858 the transition was completed and India came under the direct rule of Crown and Parliament. The Governor-General became the Queen's Viceroy, Westminster exercised direct control over the policies pursued in India and the Board of Directors was no more. The fact that in the process the Schlagintweits had lost their patron bothered no one.

There is also a hint of sour grapes about the British attitude. What, after all, were these three Teutons doing, poking around the sensitive frontiers of India? They had even managed to visit Kabul. Were there not thousands of Englishmen, skilled and brave enough, who would give their right arms for such opportunities as were afforded the Schlagintweits? It was on the recommendation of Baron Humboldt, the greatest traveller and geographer of his day, backed by the King of Prussia and the Royal Asiatic Society, that the brothers had been taken on by the Honourable Company. They were to complete a magnetic survey of India, and they arrived in Bombay in 1854. The editor of the *Bombay Times* found them well-off for instruments but none too sure of how to use them. Two of the brothers smoked so heavily he doubted if they would reach Calcutta, and he summed up the general feeling by disparaging their mission as either doomed to failure or too detailed to be completed much before the end of the century.

In London the Royal Geographical Society, who also had other preoccupations at the time—namely the squabble over the source of the Nile—paid only scant attention to their activities. And, when in 1861 the brothers offered to dedicate their great work to the Society, the offer was refused on the grounds that the Society had neither commissioned nor encouraged them. No doubt there was some resentment, too, over their recognition from an altogether unexpected quarter. For in 1859 the Tsar had conferred on Robert Schlagintweit the title of Sakunlunski—Lord of the Kun Lun. It confirmed the worst suspicions of those who had always opposed the idea of employing foreigners on such delicate missions.

The President of the Society had once described the penetration of the region north of India as a threefold process. First, and well out in front, were the reports of native travellers which shed a wide but uncertain light on the vast unknown. Behind them, piercing this gloom, came narrow shafts of clearer light representing the travels of individual European explorers. Finally, and well behind, came the zone of harsh white reality shed by the surveyors and map-makers. It was a good enough simile in the 1830s and 1840s but by the 1850s the pioneers were temporarily penned back by political and physical difficulties. The surveyors were catching up with and overtaking them.

The Schlagintweits themselves were supposedly engaged on survey work. In fact their enquiries extended to almost every branch of science and their surveying amounted to little more than not always accurate astronomical observations to fix their positions. Even this was regarded as superfluous. For in 1855 the Grand Trigonometrical Series of the Survey of India commenced operations in Kashmir. This, the Kashmir Series, was to be the crowning achievement of one of the most ambitious scientific projects undertaken in the nineteenth century.

The technicalities of map-making are daunting, but no one who has had a taste of India can fail to marvel at the magnitude of the task. How much more so is this true of the Himalayas. The ordinary traveller would usually seek out the easiest route, stick to existing tracks and climb only the unavoidable passes. The surveyor had to cover all the ground. He was after vantage points. Across empty, trackless regions he moved from one high peak to another. Mostly he travelled on foot; horses were useless. He was as much a mountaineer as a traveller, and this in the days well before aerial surveys and wire-less contact when there were no light-weight rations and no oxygen. Mountaineering skills amounted to strong lungs and a cool head and climbing equipment was what you might find in the garden shed, a spade, a pick and a hefty coil of rope. No wonder Godwin-Austen, in his old age, felt bitter about the technological dodges of the twentieth-century mountaineer.

Their instruments were delicate but heavy. The fourteen-inch theodolite used in Kashmir must have weighed about a hundred-weight. It had to be carried over rivers and glaciers and up the steep climbs suspended from a pole borne by two men. The slightest jar was liable to upset its accuracy. The plane-tables, too, were bulky affairs. Keeping the survey parties in supplies was in itself a major operation, and many a long vigil at a station in the clouds had to be abandoned because the food had run out.

With the reluctant blessing of Gulab Singh operations started from a base line just east of Jammu in spring 1855. The Grand Trigono-metrical Survey (GTS for short) had begun fifty-five years before at Madras. Its object was to construct a highly accurate framework for a map of the whole of India. The framework was to consist of meridional (north–south) and longitudinal (east–west) chains of connected tri-angles observed across the length and breadth of the subcontinent. A base line would be carefully measured between two vantage points. In the south they might be *gopurams* (gateway towers) of temples or in the plains, where no handy eminence was to be found, specially constructed towers. Perched on one of these 'trig. stations'—scaf-folding had to be wrapped round the *gopurams*—the surveyor with his ponderous theodolite measured the angle between his base line and the line to a third vantage point. By doing this from both ends of his base line, he could deduce the exact position of the third point and the distances to it, i.e. the lengths of the other two sides of his triangle.

One of these would then be selected as the next base line, the same

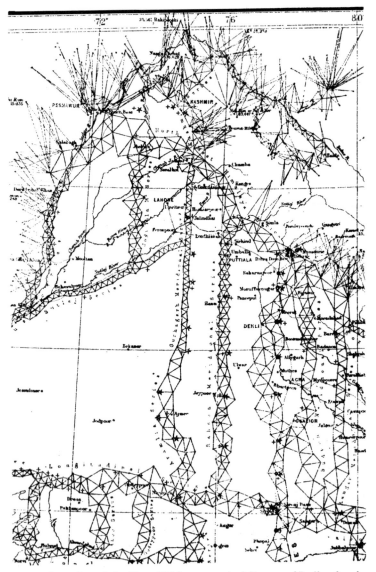

Map 4 A section of the Grand Trigonometrical Survey of India, showing
longitudinal and meridional series

process would follow and a second triangle would be constructed and so on. In practice there were many refinements and complications. Allowances had to be made for differences of altitude, the curvature of the earth and a phenomenon called refraction. Occasional checks had to be made by measuring a base line on the ground against its distance as established trigonometrically, or by tying in with another series. The accuracy was remarkable. A series carried for hundreds of miles would be found to have erred by no more than an inch or two per mile of base line.

When Vigne met the Surveyor-General, George Everest, in 1834 he was busy measuring a base line at Dehra Dun at the foot of the Central Himalayas and at the end of the principal north–south series, the Great Arc of the Meridian, which had been carried all the way from Cape Comorin. From it at Dehra Dun branched off two longitudinal series. One ran east along the base of the Himalayas into Assam. The other ran north-west through Simla and the Punjab to Peshawar. This last was finished in 1854 and, since the men employed on it had already had some mountain experience around Simla, it was they who were now deployed in Kashmir. One of the lines of their just completed series became the base line for the new series.

In the 1855 season they took their triangles across the Pir Panjal. The difficulties were appalling. Here there was no shortage of vantage points but to find a peak with a view in the desired direction, which was not blocked by another, entailed innumerable fruitless ascents. Having at last reached a suitable station, puffing and blowing from the average elevation of 15,000 feet, they had to start digging. The first job was to excavate a level platform and the best place for it was the very top. On Muli peak they dug and dug through the hard snow. What looked like the top turned out to be a cone of ice. They tried another spot and were delighted when 'only eleven feet down' they struck rock.

The advance party, whose commander was William Henry Johnson, was responsible for this selection of the stations and for the preliminary, approximate triangulation. Having dug out a platform he had to construct a masonry pillar to take the theodolite and a hut for the lampmen. They would remain there to flash their heliotropes and tend their lamps while the station was being observed from elsewhere. Lampmen would expect to be marooned for weeks, but for Johnson and his gang, living under canvas, delay was more serious. In spite of immense cold, exposure and exhaustion they had to push ahead.

They melted the snow, mixed it with lime, cemented their pillars, threw up a stone shelter for the lampmen and prayed for the clouds to lift. If they were lucky they might complete their observations in a day or so. The clouds would break. A vast array of snowclad peaks would rise from the blanket of mist in the valleys. An impossible, unearthly world of the purest beauty etched out of a deep blue sky, it seemed quite untouched by man. Then from a distant peak, indistinguishable from a hundred others, would shine out the tell-tale pinpoint of light, the 'sight never to be forgotten of a well served heliotrope'.

Alternatively the clouds might not break. Food would start to run short. The men would be suffering from headaches, snow-blindness, altitude sickness and cold. The wind would throw down their tents, the snow would make all movement dangerous and their fuel would have to be carefully conserved. Thunderstorms, which are virtually unknown in Ladakh, could be a serious hazard in the Pir Panjal. On Muli peak again, work on the platform had to be stopped when lightning set their hair and clothing crackling and great sparks leaping about. Lieutenant Montgomerie, who was in charge of the whole series and who did the final triangulation, also had trouble there. 'The small iron stove in my tent began to crackle in the most unpleasant manner . . . and the hair of my dog crackled and, in the dark, sparks were visible.' Another man had his hair set on fire and a third, who put up his umbrella to deflect the hailstones, found the thing operating as a conductor. It crackled away and 'on shutting it down it fairly hummed'.

For the winter they returned to headquarters at Dehra Dun to check their calculations and instruments, fill in the map and draw up a report. Then back again to Kashmir in the spring. In 1856 they continued up through the valley itself. One of Johnson's stations was set up on Haramukh, the presiding mountain of the Kashmir valley and there on September 10th arrived Montgomerie with the big theodolite. The ascent had taken him four days and the altitude was over 16,000 feet. Below the station stretched a fine glacier which sloped down into the sacred, trout-filled lake of Gangabal. The view to the north was the best they had yet had.

I had the pleasure to see the various ranges of the Himalaya right up to the Karakoram. There was nothing remarkable in the first six or seven ridges . . . Beyond came the snowy points of the

Karakoram range and behind them I saw two fine peaks standing very high above the general range . . . possibly 140 miles away from me.

He managed to get bearings on both these peaks and made a quick sketch of their outlines in the margin of his angle book. The larger with its double summit he designated K1, the smaller more pointed peak he marked K2. In 1856 there were no further sightings of these giants and, without another bearing on them, no way of telling their position or their height. But in 1857 and 1858 a surveyor working across the Deosai plains to Skardu again picked them out. So did Johnson in 1859. K1 was revealed as the nearer and lower of the two, about 25,600 feet. Its local name was Masherbrum.

K2 was calculated at 28,287 feet. 'The peak', wrote Montgomerie, 'may therefore be considered as the second highest in the world.'* And he stressed that the ranges north of the Indus were proving far higher than any of the earlier travellers had imagined. There appeared to be no local name for this shyest of all giants. 'Keychu', the name elicited from their Balti porters by the earnest climbers of later years, is just a variation of those other 'local' names, 'Keytoo' and 'Kaytoo'. In 1856 Peak XV in Nepal was proclaimed Mount Everest in honour of the retired Surveyor-General. Soon after, it was proposed that K2 be called Mount Waugh after his successor or Mount Albert after the Prince Consort. Neither of these names was accepted, and in 1886 further attemps were made with Mount Montgomerie and Mount Godwin-Austen. It could be argued that Johnson was probably the first to see the mountain when he erected the station on Haramukh but it was certainly Montgomerie who first recorded it. Yet Godwin-Austen was the name which came nearest to being recognised. It was, in fact, rejected by the Royal Geographical Society but a few un-official maps took it up and to this day the BBC, for example, calls it Mount Godwin-Austen.

This represents something of a personal triumph for, though Godwin-Austen was not the man to crave the immortality of having his name on the map, he had no love for Montgomerie. He accused him of gross unfairness in his summaries of the work done by the Kashmir series, of grabbing all the credit that should have gone to men like Johnson and himself. Montgomerie had 'got his honours and made a name for himself with as little personal hardship as any man I know

* Mount Everest had been recognised as the world's highest in 1852.

in the Indian Survey'. He had hardly ever gone beyond Srinagar and 'certainly never climbed higher than 15,000 feet'.

If this were true, then Montgomerie had never been up Haramukh or seen K2. But actually Godwin-Austen was quite wrong. In the first two years of the Survey Montgomerie spent most of his time in the field. In 1857 the outbreak of the Mutiny, closely followed by the death of Gulab Singh and the succession of his son Ranbir Singh, necessitated his keeping a closer eye on events in Srinagar. Added to which he had the job of instructing an endless succession of new recruits and initiating others, Godwin-Austen among them, who joined the Survey half way through. Both as organiser, political liaison and chief surveyor, Montgomerie deserved his honours, including the Royal Geographical Society's Gold Medal in 1865.

The Kashmir Series differed from most, in that triangulation and topographical work went ahead simultaneously. In other words, while Johnson and Montgomerie were busy laying down the framework of triangles, men with plane-tables who filled in the detail were hard on their heels. Lieutenant Henry Haversham Godwin-Austen* was one of these. He joined the Kashmir survey in 1857. In his first season he was the victim of another of the occupational hazards of surveyors. Near Jammu a band of irate villagers set about him and beat him up so badly that he was invalided out of the Survey and returned to England. No doubt he had trespassed on to some hill sacred to the local Hindu population. In 1860, lured as much by the charms of Pauline Chichele-Plowden as by the call of the mountains, he was back in India and back with the Kashmir survey.

Godwin-Austen was cut out for a distinguished career as a military surveyor. His father was a well-known geologist and his grandfather, under whom he had served as ADC, was a general and a KCB. He was educated at Sandhurst where he learnt topographical drawing from 'a master of the old French pictorial school' and received a 'certificate of superior qualifications'. Montgomerie could never quite get over the stylishness of his work. Repeatedly Godwin-Austen produced 'the most artistic board of the season', a somewhat laced compliment, one feels. He was also a keen geologist and had a feeling for both the texture and structure of savage terrain.

In 1860 he was immediately despatched to the bleak upper valleys of Baltistan where these skills could be put to advantage and where

---

* His father added the name Godwin, his mother's maiden name, in 1853. Until well into the 1860s Henry was known as plain Austen.

there was little danger of another brush with incensed villagers. He was perhaps not the most diplomatic of men. But he was strong, compact and immensely hardy, the ideal man for the mountains. The following year he married Miss Chichele-Plowden and then returned to Baltistan. His honeymoon was just over but, plagued by the difficulties of supporting his bride on a surveyor's pay, he was probably happy to see the scarred slopes of the Karakorams again. The view from his first stop, a trig. station on the edge of the Deosai overlooking Skardu, helped to put such things in perspective. 'Peak K2 appeared of an airy blue tint surrounded by the yellower peak K1, K3 and others all over 24,000 feet in height. Other minor peaks by hundreds thrust up their heads—some snow-capped, some rounded, some bare and angular, running up as sharp as needles.'

The triangulators might hop from one peak to the next, covering vast distances in a single season, but the plane-tablers moved much more slowly. They had to cover every glacier, every side valley and even for a ¼-inch map, which was reckoned good enough for a howling wilderness like the Karakorams, this meant a painful progress at ever-increasing altitudes. Godwin-Austen moved across the Indus and up the Shigar valley opposite Skardu. His first destination was the Mustagh pass. Thomson had listed three passes over the Karakorams. The middle one, and the only one about which anything was known, was the Karakoram pass. But east of there, from Rudok in Tibet to Kotan, Moorcroft had reported rumours of not just a pass but a road; while west above Skardu, amongst what were now proving the highest peaks of all, Vigne and Thomson had both heard of another route to Yarkand. This was the Mustagh pass.

The surveyors, of course, were supposed to be mapping the mountains, not crossing them. The Karakoram watershed was here regarded as the most likely and convenient frontier of the Kashmir state and officially no one was supposed to trespass beyond it. On the other hand they were asked to pay special attention to any passes and learn as much as they could about what lay beyond them. This was too much like an invitation to go as far as possible without getting caught and it is clear from Godwin-Austen's narrative that he intended, if possible, actually to cross the Mustagh. In fact, if he could get a small guard, he thought he could carry the survey 'into Yarkand country for a considerable distance'. The population there could not be great, 'nor their matchlocks much to be feared'.

Without a guard but with a party of sixty-six including assistants, porters, guides and a local liaison officer to cheer the men and smooth some of his more abrasive outbursts, he made straight for the pass. The way lay up the Shigar river over the Skoro pass and then along the great Panmah glacier. The only deviations were for what he usually dismisses as 'a steep pull' up some towering ridge from which he could do his mapping. Invariably this meant a climb of three or four thousand feet. To a neat and hard little man at the peak of fitness it was nothing. He swarmed up the rock faces and the icy slopes with the jerky ease of a spider traversing its web. And fretted, no doubt, as he waited for the heavy plane-table to arrive.

His sketching illustrates superbly the incredible perpendicularity of the central Karakorams. In the valleys the eye never pans. Always it is climbing. The neck aches from the unaccustomed action, and the sky is just a slit directly overhead. Even the side valleys offer no vistas. Minor glaciers fill them to a depth of hundreds of feet and, high above, loll over the lips of their moraines like malevolent tongues.

What no drawing can capture, but which his narrative emphasises so well, is the instability of it all. In the early hours of the morning when the frost grips everything it seems like a silent, timeless, unchangeable world. But in summer by mid-afternoon all hell may break loose. At one campsite he was taking an uncharacteristic siesta when he heard 'an unusual rumbling sound'. Someone shouted that the stream was coming and, seconds later, a black mass came bumbling out of a side ravine and bore down over the boulder-strewn bed of the valley like an express train. Rocks, ten feet tall, in a wall of mud and stones thirty yards wide and five deep, tumbled past 'like peas shot out of a bag'. It was a *shwa*, a stream which, jammed up for months by the snout of an advancing glacier, had suddenly broken through. The phenomenon is common enough in the Karakorams. The major tributaries of the Indus, indeed the Indus itself, have often been dammed in this way with fearful damage and loss of life when the offending landslip or glacier is finally broken by the lake behind it.

Four days later, at the foot of the Panmah glacier, he sat up late into the night listening to the blocks of ice and snow crashing down the terminal cliff of the glacier. As the cold increased, the intervals between each crash became longer until the glacier was finally lulled into sleep. Some of the glaciers were in retreat, their moraines of rock lying miles in advance of their present snouts. Others were still advancing. Glacial lakes had disappeared, old campsites had been overrun and he could see how the ice was mowing down all in its path, cutting through any low spurs, rooting up boulders, earth and scrub and bulldozing all before it. The devastation and mess created by a glacier surpass the wildest dreams of a construction engineer. Some appear sleek and white on the flank of a mountain but visit their vicinity, try to ascend them and you find them surrounded by an oozing morass of mud and stone, blocks of ice and shattered rock, as insecure and unpleasant to cross as the worst ice-field. This is the world as it must have been at the dawn of creation, cracking and crumbling, oozing and flowing with irresistible elemental forces. No place at all for human beings.

Godwin-Austen noted it all with calm wonder and pushed on for the Mustagh. Two days' march up the right side of the Panmah glacier the guides declared that they had reached the place to cross it. The width here was two and a half miles 'through as extraordinary a scene as the imagination could picture; it was the desolation of desolation'. The cloud hung low over the ice, occasionally revealing and magnifying the gaunt slope of a mountain, grey or ochre in colour, precipitous and savage in outline. The surface of the glacier consisted of stony ridges giving way to purer ice, frozen into waves and pitted with pools of deep green water. The air was full of the sound of invisible rocks crashing down slopes of ice and splashing into the pools below.

The last night before attempting the pass was spent on the ice. A few rocks were rolled together to make some sort of floor for the tents, but it was still bitterly cold. The tentless porters, without even the comfort of a fire, curled up three to a blanket but got no sleep. Even Godwin-Austen did not stir from his tent until the sun was full upon it. With eight picked men he started for the crest. The altitude and snow-glare made all movement painful but it was the crevasses which looked like being the worst problem.

They soon became more numerous and were ugly things to look into, much more so to cross—going down into darkness between walls garnished with magnificent green icicles from six to twenty feet long and of proportionate thickness, looking like rows of great teeth ready to devour one. I tried with our ropes to sound the depths of some of these fissures, but all of them tied together only made up 162 feet which was not enough. The snow lay up to the edges of the crevasses and travelling became so insecure that we had to take to the ropes, and so, like a long chain of criminals, we wound our way along. In this mode we moved much faster, each man taking his run and clearing even broad crevasses if they crossed the direction we were travelling.

In the heat of the sun the snow became soft. Men had to be sent on ahead to probe through it with long poles for hidden crevasses. Progress grew 'provokingly slow'. Finally, a mile from the top and just five hundred feet below it, the weather defeated them. The clouds descended and it came on to snow heavily. They were lucky to reach camp in safety as the blizzard howled about them and the glacier started emitting 'the most disagreeable noises—crunching, splitting and groaning to an awful extent'.

Godwin-Austen hoped to get another crack at the Mustagh but never did. Surprisingly, his failure had not convinced him that the pass was too difficult to be of commercial or strategic importance. On the contrary, he reckoned that the track up the Panmah glacier could be made suitable for pack ponies, considered the ascent to the pass encouragingly gradual and reported that it was in regular use. The worst part of it was not the physical difficulties but the danger of being attacked by robbers on the other side. It looks as if the government took his idea of what was practicable with a pinch of salt, for it was not till 1887 that another attempt was made to cross the Mustagh. This was by Younghusband, and largely on his own initiative. He succeeded, just, but corroborated what Vigne and Thomson had said; namely that the pass had once been much used but, early in the century, the sort of glacial movement which Godwin-Austen describes so vividly had virtually closed it to all but mountaineers.

The old pass had debouched not on to the Panmah but on to the Baltoro glacier for which Godwin-Austen now made tracks. He was not finished with the Karakorams by a long chalk and arguably his ascent of the Baltoro was a greater achievement than his attempt at the Mustagh. From it stems his association with K2.

For the meeting of the Royal Geographical Society at which his paper on 'The Glaciers of the Mustakh [Karakoram] Range' was read, his father, old Godwin-Austen the geologist, came up from Guildford. On the way he tried to translate young Henry's extraordinary discoveries into terms everyone might appreciate, and after the paper was read he rose, the proud father, to elucidate. ' If Hampstead and Highgate were presumed to be high mountains, these glaciers would descend as far south as Tunbridge and north two-thirds of the way to Cambridge.' It was a brave attempt, but no one has ever succeeded in conveying to those who have not known the Himalayas any conception of their size. To talk of mountains nearly ten times the size of Snowdon or twice the height of Mont Blanc means nothing. Homely comparisons only belittle the grandeur and no attempt at a scale can convey the effect of such mighty scenery. It is not even enough to have seen it. With head throbbing, lungs pulling fiercely at the harsh liquid air and feet burning with blisters, one must feel it. And, at the same time one must experience that very different, wholly beautiful sensation which is the peculiar reward for penetrating the greatest mountains. An unburdening of the spirit, an ennobling really, it too is indescribable; to know it, one must feel it.

If this experience intensifies with the altitude the mountaineer is to be envied. But few mountaineers can ever encounter such grandeur as awaited Godwin-Austen up the Baltoro glacier. For this is the innermost sanctum of the Western Himalayas, an amphitheatre of the greatest mountains on this planet. In the space of about fifteen miles the Baltoro holds in its icy embrace ten of the world's thirty highest peaks. They line its sides and close its easternmost end like high priests guarding the Holiest of Holies. Sir Martin Conway who, thirty years later, was the next to enter this great nave, aptly named the peak at the eastern end The Golden Throne.

Up this broad aisle of ice, mere specks on its noble extent, Godwin-Austen and his men now made their painful progress. Others might have felt self-conscious and awed into reverence by such surroundings. But not Godwin-Austen. He had his share of the calm confidence of men like Henry Lawrence plus something more. The next generation would be inclined to romanticise and to dwell on the wonders of Nature with a capital 'N'. But the men of mid-century spurned such nonsense. This was a period of technological revolution. Railways were changing the face of India, steamships had halved the time it took to get from Southampton to Bombay, and the telegraph was changing the whole character of government. Such developments had more effect on the administration within India, and on the British government's involvement in the country, than did the demise of the Honourable Company. The GTS was at the head of these advances. Their maps were essential for the planners of railways and telegraph lines; and, if this sounds fanciful in the case of the Himalayas, it should be noted that already a railway into Kashmir was being considered, and within twenty years there would be talk of a line to Kashgar. The men of mid-century, especially those involved in these advances, reverenced facts and figures. In the Karakorams it was the ice four or five hundred feet deep, glaciers over thirty miles long, peaks up to 28,000 feet, that mattered.

Godwin-Austen got quietly on with his job in the most practical way possible. His main problem was to discover where K2 lay in relation to the Karakoram watershed. Often the highest mountains lay not on the main watershed but on a spur from that range. It could be south of the watershed but, from Montgomerie's description, it looked as if it was north of it and therefore beyond the frontier. From the glacier he could see Gasherbrum straight ahead and Masherbrum towering to the right. To the left, where K2 should lie, intervening spurs cut off the view. In the hope of a sighting he started to scale Masherbrum. A thousand feet above the glacier there was still no sign of it. Two thousand feet up and he thought he could distinguish a more distant lump of rock and snow just nosing above the horizon. 'After another sharp push up to a point from which it was impossible to mount further there no longer remained a doubt about it. There with not a particle of cloud to hide it it stood the great peak K2.' It was actually on the watershed.

So, though Godwin-Austen had not discovered it, it was he who first saw K2 at close quarters and ascertained its position in relation to the Karakoram system. If the peak itself is not strictly speaking his, there is at least a Godwin-Austen Glacier. Fittingly it is the one that seemingly pushes the watershed northwards so that it actually runs through K2.

*   *   *

Godwin-Austen had sixteen brothers and sisters, four of whom lived to be over ninety and two to be over a hundred. He himself reached eighty-nine and, when he died in 1924, he was revered as something of a grand old man of the mountains. At seventy-three he published *The Fauna of British India*, vol 1, at seventy-five he was awarded the Royal Geographical Society's Gold Medal and to the day of his death was still a forceful character with strong views on the rights and wrongs of mountaineering. He came to be regarded as the greatest of the Kashmir surveyors and the first mountaineer in the Western Himalayas.

It is an invidious business singling out a couple of men from the dozen or so who worked on the GTS in Kashmir. They all performed feats of endurance and nearly all visited hitherto unknown regions. But none more deserved a share of the acclaim given to Godwin-Austen, and none was less recompensed for his efforts than Johnson. His name rings no bells. He won no medals and he died, or more probably was murdered, in obscure and ignominious circumstances. Yet Johnson was the man who, excluding the Schlagintweits, first reached Eastern Turkestan.

In the early days of the Kashmir survey he was Montgomerie's right-hand man. His energy was prodigious, his work as a triangulator brilliant and his courage as a mountaineer unsurpassed. Year after year he broke the altitude record, 19,600 feet, 19,900 feet, 20,600 feet, and in 1862 he built a station at 21,000 feet and climbed to the, for those days, incredible height of 22,300 feet. Four of his stations were the world's highest for another sixty years. Besides doing the pioneer triangulation he was often entrusted with the principal triangulation and, in Montgomerie's absence, he took temporary charge of the whole Series.

In 1864, with the work almost finished, Montgomerie went on leave. A Lieutenant Carter was appointed to take his place. Johnson was furious. He had been with the Kashmir survey throughout; Carter had only just joined it. And he had been highly regarded by Montgomerie who repeatedly brought his services to the attention of the Surveyor-General. He protested bitterly but without success, and from this time onwards his whole attitude seems to have changed.

Johnson's trouble was that he did not quite belong. He was 'native born'; his parents were as English as anyone's, but his father was a mere Ordnance Conductor and young William Henry had been brought up in India. He had never been to England. Instead of Sandhurst he was educated at Mussoorie, a place where, in Emily Eden's patronising words, 'parents who are too poor to send their children home, send them'. He joined the Survey as a Civil Junior Sub-Assistant and, though he rose rapidly, his uncovenanted status could never compete with that of the young Sandhurst officers like Godwin-Austen. It was a rule that no civilian, however senior, could ever have charge over military officers.

With Montgomerie's departure and the winding up of the Kashmir series there was also a good deal of reorganisation. Triangulators were no longer needed, and Johnson was pressed to switch to the topographical side. He had used a plane-table before but the work did not suit him. He liked to be at the head of things, pioneering way out in front, and he refused the transfer.

Thus, embittered and with his career in crisis, Johnson had spent the last season, 1864, rounding off his work in the north-eastern corner of Ladakh. In the process he had again beaten his own world altitude record by reaching a point 23,000 feet above sea level. He had also crossed the Karakoram pass and continued for three days towards Yarkand before turning back. At the time the Maharaja of Kashmir, without apparently consulting the British, had established an advance garrison well beyond the pass at a place called Shahidulla. This excursion of Johnson's was therefore safe enough and though his survey work there was rather haphazard, he was not censured for crossing the frontier.

At the end of the season the Kashmir series was declared finished. There was no more work to do within the Maharaja's territories.* Yet Johnson was not done. It was as if he could not bear to tear himself away from the wild scenery where he had achieved so much and from the association with Montgomerie that had worked so well. As a parting present Montgomerie agreed to recommend that he return to Ladakh once again. Surprisingly the request was granted, and in July 1865 Johnson reached Leh alone, the last relic of the Kashmir survey.

In the furore that was to greet his exploits of 1865, the argument raged over whether or not he was authorised to go beyond Kashmir territory. His instructions did not specifically say so, and it could be argued that all the points that he was supposed to be fixing, though beyond the frontier, were to be observed from within it. This, however, was not the spirit of his instructions. In the direction of Rudok, which place Godwin-Austen, uncensured, had almost reached in in 1863, it was clearly understood that he would have to cross the frontier. From this, he argued, he was entitled to assume that the Superintendent of the Survey had obtained permission for him to go beyond the Maharaja's domains, though he was stretching credibility when he inferred that it amounted to 'a roving commission to explore Central Asia'.

A more difficult question that has to be decided is not whether he was authorised to cross the frontier but why he did so. Was it as a devil-take-them-all protest against being repeatedly passed over? Was

* Except beyond the Indus towards Gilgit. This region was still virtually unknown to Europeans. The Kashmir government was struggling to maintain its toe-hold in Gilgit itself and no amount of pressure could persuade the Maharaja to allow surveyors to visit it.

it, as he claimed, because he believed that the government would welcome such a spirited attempt to extend the boundaries of geographical knowledge and thus at last acknowledge his services? Or was the reason more political? Was he trying to force the government's hand with regard to trans-frontier policy, and perhaps even not working for the government of India at all but for the Kashmir government? A good case can be made out for each.

The bazaar, Leh

He left Leh with fifty porters, three native assistants, five mules and six horses. They headed east to the Changchenmo valley, then north across the desolate plains of Aksai Chin, and reached the Karakash river inside three weeks. This route Johnson already knew. It had the advantage of circumventing the dramatic passes of the Karakoram route but, as later travellers discovered, it was questionable whether the total absence of grazing and the continuous altitude of between 15,000 and 18,000 feet were not greater drawbacks.

The Karakash river divides the Aksai Chin plateau from the Kun Lun mountains to the north. Johnson now climbed the nearest peaks hoping to get a glimpse of Yarkand or Kotan. To have been able to triangulate their positions from the established positions of some of the Kun Lun peaks would have been a cartographical coup. But alas, there was no hope. The view from the heights of the Kun Lun was as disappointing as that which Thomson had found on the Karakoram pass. Johnson was standing not on the saddle of some simple range but on the outermost heights of another colossal mountain system. It was like the Pir Panjal; open plains lay to the south and the north was still a mass of mountains.

He should now, like Thomson, have turned back. On his own reckoning, which was extremely generous to the Kashmir government, the frontier here followed the line of the Kun Lun. To go further would mean trespassing on Chinese territory. But at this point in his narrative—and it may be significant that he leaves it till now— Johnson introduces the unexpected information that he held an open invitation to visit Kotan. The ruler of that place had heard of his travels of the previous year and had sent an emissary to Leh with the invitation. Johnson had brought the man along and now, deciding to take up the offer, sent him on ahead to ask for guides and a safe conduct. Three weeks later these were provided, he crossed the Kun Lun by an unidentified pass and reached Kotan on October 2nd.

Johnson was the first European to visit the city since Marco Polo and Benedict de Goes. Even Gardiner never claimed to have been there, and the Schlagintweits, though they reached its vicinity, never dared to enter the city. Johnson reported at length on all he saw and learnt. He was well received and, though detained longer than he wished, was allowed to depart after only sixteen days. Compared to the difficulties experienced by the next European visitors to Eastern Turkestan this says a lot for his tact as well as for the disposition of the local ruler. From Kotan he headed west towards Yarkand. If he was to get back to India before the passes were closed for the winter there was scarcely time for a visit to Yarkand as well. Yet he had good reason for trying. An offer of £3,000 had been made to him by an influential faction if he would accept the governorship of the city in the name of the British government. The city had a large Kashmiri population who could only benefit from such a development. It was probably a genuine offer, but the circumstances that prompted it, civil war within and invaders without, were enough to dissuade

Johnson from proceeding. Only thirty-six miles from the city's gates he turned back. He knew of the Schlagintweits and of the fate of Adolph who had blundered into a similar situation in 1857.

But though he had not been to Yarkand, he now returned to Leh by the classic Leh–Yarkand track. His was the first account by a European of the whole route, and he was the first to cross all five of the great passes. Coming from the north the last three, Karakoram, Sasser and Khardung, had all been crossed by Thomson and by himself in 1864. But the two Kun Lun passes, the Sanju and the Suget, were new to geography. Johnson scarcely does them justice. His idea of a difficult route is almost unthinkable. He could now claim to be the most experienced mountain traveller of his day and anywhere that boasted a defined track was like a bowling green to him. On the way out in mid-summer he had noted, as a curiosity rather than as any indication of the hardships, that the icicles in his mane-like beard had failed to melt even in the sun. What the return journey in October and November must have been like it would be hard to imagine without the accounts of later travellers.

In 1889 Dr. (of Divinity) Henry Lansdell also crossed the Sanju and Suget passes in October. The frost had its advantages. It made the rivers fordable, the footing firm and reduced the risk of avalanches. But the cold brought its own set of problems. Keeping a journal or, as Johnson found, plotting a map, was impossible because the ink froze between the heated pot and the paper. A cup of coffee not downed at the first sip, froze solid in minutes and the apple at Lansdell's bedside was a rock by morning. On the Suget, a month earlier than Johnson, he nearly suffocated from the need to wear his full wardrobe.

> To begin with I had put on a thick lamb's wool vest with sleeves and drawers, then ditto of chamois leather; next a flannel shirt and above it a chamois vest without sleeves lined with flannel; cloth trousers and waistcoat, with jacket of kid leather, flannel lined; then an ulster lined with fur; and above it for sleeping my Khoten coat of sheepskin, with thick stockings and fur lined boots, together with a woollen helmet for a nightcap. Thus I lay down on my four trunks while Joseph covered me with shawl and lambskin. This represented my maximum—namely five skins besides my own, four flannels and a thick coat; yet with all this at Suget it was cold and I never got into a perspiration, though the weight of clothing

and the effect of the *dam* [altitude sickness] proved a little too much; and I had to rise in the night feeling half suffocated . . .

It was while Johnson, less generously clad, was negotiating the Kun Lun passes that the storm broke back in India. First news of his dealings with Kotan reached the Lieutenant-Governor of the Punjab from the Maharaja of Kashmir. A letter from Johnson to the Super-intendent of the GTS explaining why he was off to Kotan arrived soon after. But the first mistake had been made. As in the case of Moorcroft's dealings with the Ladakhis, it was regrettable when news of such ventures reached the government through a third party. They looked stupid not knowing the movements of their own people and were inclined to issue a harsh repudiation. In Johnson's case the GTS took the first brunt. The Superintendent was reminded that his officers had no right to cross the frontier without government's per-mission. He replied claiming that Johnson had acted on his own initiative, and then duly censured him. But he also entered a strong plea in his favour. No man was more conscientious in his work or more successful in his dealings with native dignitaries. He would not have gone without first assuring himself that he was welcome; the dangers he was likely to encounter were more physical than political. The government hoped the Superintendent was right, and decided to postpone all further action until Johnson returned and his report had been submitted.

This was ready by the following April. As soon as it was published the geographical world rose in Johnson's defence. In London, Sir Roderick Murchison of the Royal Geographical Society declared that he had never read a paper that better exemplified 'the character of a true, bold and scientific manager of an expedition' (and this just a few months after Sir Samuel Baker, for his explorations up the Nile, had shared with Montgomerie the Society's honours for 1865). Lord Strangford seconded him, calling it 'one of the most important papers that had ever been read before the Society'. All roundly condemned the government's treatment of Johnson and, by inference, the cautious isolationist policy of the then Viceroy. In Calcutta, Colonel Walker, the GTS Superintendent, in words that he was soon to regret, told the Asiatic Society that Johnson's paper was 'the most valuable con-tribution to the geography of Central Asia that has been made for several years by anybody in India'.

The government, too, were pleasantly surprised by the report.

Johnson's political and commercial findings were every bit as important as his geographical labours. For nearly ten years there had been rumours reaching Leh of momentous goings-on beyond the mountains. Johnson now confirmed them. Chinese Turkestan was no longer Chinese. The predominantly Mohammedan population had massacred their Chinese overlords and the whole country was now in a disturbed and vulnerable state. Kotan itself had reverted to the status of an independent city-state and was ruled by an ill-tempered but apparently friendly octogenarian. Fearing for his hard won and tenuous independence, he had entertained Johnson in the hope of getting British assistance and support. The danger was all too evident in the affairs of Yarkand where an invading army from Khokand waited at the gates while inside the city the recently liberated population fought amongst themselves. Johnson thought a consignment of arms and a few native sepoys to act as instructors would stabilise matters in Kotan, but neither to the octogenarian nor to the factions in Yarkand did he make any firm commitment. In the government's view this was greatly to his credit.

On the commercial front Johnson painted the picture of a populous land, rich in minerals, that had suddenly been deprived of its one, and almost its only, trading partner. The Chinese had gone and with them the market for Kotan's gold, her jade and her skins. Desperately missed too was the compressed tea which had been imported from China; any country able to make good the deficiency was sure of a warm welcome. And who better than British India? For though Johnson's outward route over the Kun Lun was unthinkable for trading purposes, he confirmed the existence of a road 'suitable for wheeled carriages' from Rudok to Kotan. This was the same that Moorcroft had heard tell of and which Thomson had listed as the third of his routes into Eastern Turkestan. There was said to be grazing and fuel at every stage and only the Tibetans to be bribed before throwing it open to Indian tea caravans.

In subsequent years many lives were risked, and some even lost, looking for this supposed highway of commerce. Johnson's optimism also seems to have ignored the little matter of Tibet still being in Chinese hands. Neither they nor the Tibetans were any more disposed towards new trading patterns than they had been in 1847. But the one note of caution that he did sound was a warning more calculated to rouse the authorities in India. In trade, as in politics, Eastern Turkestan stood wide open. And the Russians were already stepping into the vacuum. He suspected that the invaders from Khokand who were now besieging Yarkand were the precursors of direct Russian intervention. He met a Jew in Kotan who admitted to being an agent of the Russian government and he reported that Russian caravans were already regularly penetrating as far as Kotan.

Sir John Lawrence, the Viceroy and brother of Sir Henry, conceded that Johnson had made the most of his opportunity. He should not have crossed the frontier but he had already been reprimanded for this and, since he had done so, it was 'satisfactory' that he had acted so creditably. He was prepared to pay Johnson's expenses and take no further action. Another Viceroy might have offered him promotion, a reward or at least official congratulations. But not Lawrence. 'Masterly Inactivity' was how his foreign policy was usually characterised. Many questioned the 'masterly', few the 'inactivity'. He was not interested in Central Asia and would not willingly condone, let alone encourage, transfrontier exploits.

This was no good to Johnson. He was the kind of man who threw himself wholeheartedly into everything. He loved nothing more than a *tamasha*, a handy word that can mean any kind of excitement from a party to a brawl. In this case it was the latter. Johnson went to see Lawrence. Burning with righteous indignation he stressed his long record of service, cited the endless reports in which he had been commended and showed how four times he had been unfairly passed over. Next he contended that his instructions for 1865 clearly implied that he was expected to cross the frontier. Thirdly, and this was entirely new, he declared that he had not gone to Kotan of his own free will. On reaching the Karakash river he had found 200 men lying in wait for him. They had compelled him to go to Kotan. He did not claim that he went reluctantly or that he was not already flirting with the idea. But the way in which in his first narrative he withheld any mention of Kotan till this stage in his journey does suggest that there was some important development here. As we will see, the government at any rate were prepared to accept this new version.

A lot more followed. The reason he had waited till now to reveal the true nature of his visit was because previously he had been trusting to the government's voting him a substantial reward for his initiative. With this out of the question, his only hope of liquidating the heavy debts he had incurred when buying his release from the grasping officials in Kotan lay in his telling the full story and asking the government to make a compassionate settlement. The amount in question was £1,600. He reckoned the value of the surveying he had done en route at more than five times that, besides which he claimed compensation for 'the enlargement of his heart' caused by continuous exposure to the effects of high altitude. And so on.

Lawrence told him to put it all in writing. Within four days of his doing so Colonel Walker, the Superintendent of the GTS, hit back. This time there was to be no standing up for his subordinate. He had done his best for Johnson, but now the man was clearly out to make trouble and to embarrass the service. The gloves came off and the slanging match that went on well into 1867 does little credit to either party. Distasteful though it was, Walker felt bound to draw the government's attention to Johnson's having neither 'the benefit of a liberal education' nor the ability 'to rise above the disadvantages of his position'. His report had had to be entirely rewritten, his map recast and his observations re-reduced. None in their original state was fit for publication. Johnson denied all this. The errors in his survey had crept in at the Surveyor-General's office. Walker was now trying to denigrate his journey and to disclaim all responsibility. And he did so because he now realised that it was not to be a case for congratulation. But in the beginning, hoping to cream off some of the expected credit, he had been happy enough to defend Johnson's action. Why else had he written of the journey in such glowing terms?

Two points of more substance came out. One was that Johnson, before leaving his home in Mussoorie in 1865, had raised a considerable sum of money as if in expectation of heavy expenses in the mountains. The other was that by now, 1867, he had accepted a post with the Kashmir government on three times the salary he had been receiving with the GTS. Combined, these two bits of evidence cast a sinister light over the whole affair. It looked as if he could have planned on going to Kotan as early as 1864 when he put Montgomerie up to recommending the last visit to Ladakh. Had he, all along, been in collusion with the Kashmir government? The Maharaja was known to harbour designs on the cities across the mountains. His father's expansionist policies were not forgotten. There was the advance garrison at Shahidulla which, in 1864, had led Johnson to claim that the Kashmir frontier lay some hundred miles further north than most maps showed it. And in 1865, like a runner breasting the tape, Johnson again pushed it forward, this time to the Kun Lun peaks north of the Karakash river. Even if not at the time in collusion with the Kashmir government, he could have been angling for some sort of subsequent recognition from them.

On the other hand, his activities could all be interpreted as a loyal, if over-zealous and misplaced, attempt to bring the frontier of British India nearer Eastern Turkestan and thus involve it in the affairs of Yarkand and Kotan. His report was not uncritical of the Kashmir authorities and his commercial recommendations were clearly of advantage only to the British. He had always been well liked by the Kashmiris and the Ladakhis and, heaven knows, he had had ample provocation to seek employment outside the Survey.

In 1872, after five years of obscurity, Johnson emerged as the Maharaja's wazir, or governor, of Ladakh. His predecessor had also been an Englishman but it was still a considerable achievement. His rule was distinguished by 'all kinds of *nachs* and *tamashas*', but he ruled fairly and openly and was well loved. Moorcroft would surely have approved. He worked for the improvement of cultivation and trade, and dispensed justice with the speed and directness that a simple people like the Ladakhis appreciated. He also maintained excellent relations with the British government and continued to show a great interest in the opening of the road to Yarkand. Partly in recognition of this, and on the commendation of his old friend Montgomerie, the Royal Geographical Society presented him with a gold watch in 1875. In the same year he made his one and only visit to England. Eight years later he was removed from office and died soon after. He had probably been poisoned. It was one of the occupational hazards of working in the intrigue-ridden atmosphere of the Kashmir court and had no apparent bearing on the Kotan affair.

Godwin-Austen wrote a generous obituary of him. He seems to have taken the view, which in the end is probably the right one, that the political undertones of Johnson's journey to Kotan were incidental. He was not, at bottom, an agent for anyone but just a very disappointed careerist. The government too seems to have taken this view. They accepted the later version of his story and even agreed to refund the £1,600 spent on bribes to secure his release. As for the money he had raised in Mussoorie, it was accepted that this was just a security measure necessary because, in 1865, there were no other GTS personnel in Ladakh and he could therefore not afford to run out of funds.

This all seems to agree with what can be learnt of his character. He was not as ill-educated or incapable as Walker tried to make out. If anything he was too able. But to someone with as undistinguished a background as Johnson's such ability only brought frustration. If it had been Godwin-Austen who had gone to Kotan, a political motive might have to be found. But Johnson had reason enough of his own. He had been passed over too often. His career was in the doldrums and he grabbed at the one opportunity which, for better or for worse, might lift it out of them. It was a desperate gamble but, to the ever zestful Johnson, all the more appealing for that.

12. In Contradistinction

In the late 1860s and early 1870s the story of the penetration of the Western Himalayas suddenly changes character. The tempo increases. The various themes of the Moorcroft overture, developed by strategists like Burnes, scientists like Jacquemont and Thomson and doubtful 'politicals' like Vigne and Johnson, at last coalesce. The crescendo of the Kashmir Survey ushers in a grand finale of which the opening of relations with Eastern Turkestan is the object and climax. The brief heyday of exploration in the Western Himalayas has been reached. Between 1868 and 1875 no corner of the unknown aroused greater interest or was tackled with more energy. It perhaps lacked some of the popular appeal of African exploration; it was less of a free-for-all and, where possible, progress was still shrouded in secrecy. But to those in the know there was no question of its significance. 'The great mountain backbone to the north-west of our Indian Empire and Eastern Turkestan' was where, according to the president of the Royal Geographical Society, all the main geographical advances were being made. The journals of the period are thick with contributions from travellers in the region. Even mass market publications like *Macmillan's Magazine* and the *Illustrated London News* carried long stories about the bazaars of Yarkand and the jade mines of the Kun Lun. The explorers themselves reflect some of this heightened interest. They are sponsored by a geographical society, reporting to a newspaper or writing a bestseller about their travels. They are a different type of man from Moorcroft, Vigne and co., narrower in their interests, more dogmatic about their discoveries and more competitive about them. Less attractive, certainly, but also more dramatic.

An unmapped region is a great attraction to the explorer but, seemingly, a partially mapped one is even more of a draw. The Kashmir Survey produced elegant maps of Ladakh, Baltistan and all the country to the south. But to the north they showed an inviting blank. The detailed contours and shading abruptly ceased, there was a narrow limbo of Kun Lun peaks and dotted rivers, then all was white, irresistible space. The men to whom it appealed set themselves specific targets. There were to be no more of those rambling aimless journeys like the travels of Moorcroft, Gardiner and Thomson. The new explorers had their sights firmly fixed on one or more of the few remaining unknowns; there was the chance of unravelling the hydrography of the Pamirs, begun by Wood but neglected ever since, or that of uniting a survey carried up from the GTS of India with that of the Russian survey department which now reached the Tian Shan. There was the question of the extent of the Kun Lun and of how rivers, draining north from the Karakorams, broke through this further barrier. And, of course, there were the twin cities of Yarkand and Kashgar. Johnson had notched up Kotan, but Yarkand and Kashgar had yet to be visited by someone who would live to tell the tale. Now, in the late 1860s, with the Chinese gone and, with them, the policy which had kept the whole country firmly closed to foreigners, a start could be made with a determined bid to reach these fabled cities.

So at least it seemed to Robert Shaw, a tea-planter from Kangra. With a caravan of merchandise and high hopes of 'opening up Central Asia', he set off from Leh on September 20th 1868. And so too it seemed to George Hayward, a professional explorer from England, who arrived in Leh the following evening. It was pure coincidence that both men were tackling the same journey in the same season, but from now on there was to be precious little coincidence in their activities. Shaw had spent two months in Leh organising his caravan and preparing the ground ahead. Hayward, finding the trail still warm, was in and out of the place in eight days. Following hard on Shaw's heels, he was within striking distance before they reached the Karakorams.

Shaw was taking things easy. For want of carriage he had left half his goods to be sent on from Leh; a slight delay might enable them to catch up. It would also give the man he had sent ahead more time to predispose the Yarkand authorities to his visit. So far as he knew he was the only one in the field. Various attempts had been made to dissuade him from going, most notably by Douglas Forsyth, a Commissioner in the Punjab who had made Central Asian affairs his speciality. In fact Forsyth was the man who had originally given Shaw the idea of undertaking the journey. He was also his most likely rival; no one more deserved the honour of opening contacts with Yarkand. But equally, so long as the cautious Lawrence was Viceroy and so long as Forsyth remained a servant of the Crown, he had no hope of being allowed to. When, in Leh, he made a final bid to warn Shaw off, the latter paid no heed; he put it down to sour grapes. There was also a Mr. Thorp who asked if he might accompany him. He was quickly disillusioned. As for the unknown Hayward, his approach was no more than a rumour. Shaw found it hard to believe that Forsyth and the Punjab authorities could have forgotten to tell him that he had a serious rival, but, just in case, he dashed off a sharp note to the interloper. If there was any truth in the rumour, Hayward was to change his plans immediately. He, Shaw, had been preparing for this journey for years; if some bungling globetrotter were now to upset the delicate state of his negotiations with the Yarkand authorities it would foul the whole future of Indo-Turkestan relations.

Nineteen days out of Leh, Shaw was in the Changchenmo valley hunting wild yak. On the way back to camp he spotted six gigantic sheep of the Ovis Ammon variety. Fine males they were, almost as big as the Marco Polo sheep of the Pamirs and, even in those days, the rarest of Himalayan trophies. A long wait followed, then an arduous stalk. He was just getting within range when the beasts suddenly bolted. Stark against the snow, in full view of his quarry stood the culprit, a stranger. In a foul temper he bore down on the man. It was not Hayward, but it was Hayward's messenger—an unfortunate moment to arrive. Nothing in their subsequent dealings led Shaw to regret the rage in which he first learnt that he had a companion.

The two men met a few days later. They dined together crouched over Hayward's campfire. It was probably a modest meal. Hayward was travelling light, in native disguise and without even the luxury of a tent. Perhaps Shaw contributed a course from his own more elaborate commissariat. A contemporary traveller in Ladakh recommends 'cubes of soupe a l'ognon au gras from Chollet et Cie of Paris, tins of hotch-potch, tins of half boiled bacon and tins of salmon for breakfast'. Shaw might also have provided his collapsible table and chairs. Together with a cane commode, collapsible bed and a tin bath, these were regarded as essential camp furniture. There are pages of such advice in every nineteenth century account of Kashmir. Camped in a place like the Changchenmo, Shaw and Hayward must have recalled them with a bitter feeling of their irrelevance. This valley is as bleak a spot as any south of the Karakorams. The ground is stones, beyond that rock, beyond that mountains. Nothing else. The only colours are the blue of the sky and the searing yellows,

19 Crossing glaciers in the high Himalayas

20 George Hayward

browns and purples of the rocks. All that moves is the biting wind. It is as dead as the moon. Yet, for Shaw and Hayward, ahead lay worse. Here the famished ponies could tear at the few widely spaced blades of desiccated grass and the travellers could make from some carefully exhumed roots a spluttering fire such as they now crouched over. In the Aksai Chin they would remember these as luxuries.

Already, in October and at a modest 16,000 feet, it was bitterly cold. On the pass into the valley Hayward had measured twenty-nine degrees of frost. Shaw's stock of claret was freezing in the bottles and bursting them—not a total disaster; the glass could still be knocked off and the contents duly passed round, each man chipping a piece into his tumbler and quietly sucking it. Thus engaged, the two men spoke their minds, and in due course reached an understanding. First Hayward would forgo his disguise. He was dressed as a Pathan and travelling in suitably spartan style. The idea had surely come from Gardiner, now roaming the bazaars of Srinagar and far from reluctant about imparting the gems of his extraordinary life to impressionable young travellers. Times, however, had changed since his day. Adventurers of doubtful origin were now regarded with intense suspicion. Moreover Hayward's experience, though it seems to have included a spell on the Afghan frontier, was not up to the disguise. Shaw rightly felt that if the first Yarkandis they met saw through it, they would both be turned back. If it held good till they reached Yarkand, the many Pathans there would soon suspect his faltering Pushtu and then their very lives would be in danger.

Secondly, Hayward was to give Shaw ten days' start. He was to mark time in the Changchenmo while Shaw sped on to the frontier and made contact with the first Yarkandi outpost. Shaw argued that if they arrived simultaneously the Yarkandis were bound to get suspicious; assuming that his agent had done his job, they would be expecting only one Englishman. The arrival of two would be sufficient excuse for refusing entry to either. On the other hand, and this was the third point, if all went well, Shaw was to do his best to warn the Yarkandis of Hayward's approach and to persuade them that he too should be admitted.

With polite wishes for a rendezvous in the palaces of Yarkand they parted. It was their last meeting for eight months. Throughout the whole of their visit to Turkestan they would never speak again. They would be the only Europeans in a vast and strange land, often they would be in fear for their lives, sometimes they would be actually within earshot of each other, yet, other than a few bitter notes, they would hold no communication. Each advanced a different explanation for this, but the conclusion must be that basically they resented one another. It is the first and only instance of the acrimony which is so notorious a feature of African exploration showing itself in the Himalayas.

There was, of course, a great deal at stake. For years Yarkand had been reckoned the juiciest plum awaiting the Himalayan traveller. It was the largest and richest of the cities of Eastern Turkestan, it was surrounded with that aura of notoriety so vividly evoked by Burnes and, in 1868, its very position was still a matter of conjecture; estimates varied by as much as two hundred miles. How much of its glamour had survived the overthrow of the Chinese was one of the many questions to be answered but Shaw, a diligent and dedicated sort of fellow, was not much exercised by this. He preferred to restrict his vision to the commercial potential of the city. He saw it as the 'Eldorado of Asia'; his job, remember, was growing tea. At school at Marlborough he had collapsed during the entrance examination for Sandhurst. He still came out top but the long illness that followed —it was diagnosed as rheumatic fever—put paid to his military ambitions. A quiet, unexciting life free from physical and mental strain was recommended. He went up to Cambridge and then out to India where the gentle foothills climate of Kangra was expected to suit his condition. Tea is still grown in Kangra though the plantations are nothing like as neat or extensive as those in the Eastern Himalayas. For one thing, it is much too far from any Indian seaport. But it might well have turned out to be quite as big business as Darjeeling or Assam. For, as Moorcroft had conjectured and as Shaw now passionately preached, Kangra was the nearest tea-growing area to the great tea-drinking population of Central Asia. And if the caravans from China were no longer reaching the markets of Turkestan, then now was the time for the caravans from India to take over.

To this extent Johnson's report on Kotan had not gone unheeded. Lawrence and his advisers in Calcutta might not be prepared to follow up the changing state of affairs beyond the mountains but in the Punjab, now a fully fledged province of British India, they were abuzz with the possibility of opening relations with Eastern Turkestan. As early as 1861, the Lieutenant-Governor of the Punjab had commissioned a report on the trade and resources of the countries beyond the frontier. Two years later the first of several attempts was made to get the Maharaja of Kashmir to reduce the exhorbitant tariffs levied on trade to and from Yarkand through Leh. In 1866 the Lieutenant-Governor was all for sending Johnson back to Kotan to continue his good work or, failing that, to let Forsyth go. The latter was then Commissioner for an area which included Kangra, Kulu and Lahul, all of which stood to benefit greatly by the trade.

Neither then, nor in the following two years, would Lawrence hear of Forsyth's undertaking such a mission but the persistent Commissioner continued to make the running and, in 1867, scored two notable successes. On his strong recommendation, Lawrence agreed to the posting of a British representative in Leh. Another of Moorcroft's dreams was coming true. The agent's job would be to enquire further into the Maharaja's supposed stranglehold on the existing trade; to soften the pill for the Kashmir government, the man chosen was a doctor, Henry Cayley. Forsyth's other success was the inauguration of an annual trade fair at Palampur, only a few miles from Kangra, to which all those engaged in the existing trade with Yarkand were invited. Not many Yarkandis ever reached Palampur. It was too far for them to go if they wanted to return across the mountains in the same season. But it did provide a useful rallying point for the Turkestan lobby in India. There, in November 1867, Forsyth announced to the assembled company that the Leh agency was to be 'permanent, at all events for some time to come'.

This 'Delphic announcement'* was jumping the gun as far as Lawrence was concerned but, in the light of Cayley's successful first year, it was a safe prediction. Trade had increased dramatically, and the agent was convinced that, if an easier route could be found avoiding the Karakoram passes, there was no limit to its possibilities. The classic route of the five great passes was direct enough, but it was never open much before June and was too dangerous to invite exploitation. The loss of carriage beasts, evidenced by that grizzly trail of bones, was reason enough. For every laden pony the wise merchant took two spares; and this in spite of, the fact that for at least three of the five passes the merchandise had to be transferred to hired yaks.

Early in 1868 Cayley reconnoitred a route further east which was roughly that followed by Johnson on his outward journey to the

---

* The phrase is from G. J. Alder's *British India's Northern Frontier 1865-95*, than which there is no better analysis of British policy towards Eastern Turkestan.

Karakash river. It was not a way round the mountains as its supporters later contended. The Chang La took the place of the Khardung pass, the Marsimik of the Sasser, the Changlung of the Karakoram and so on. If anything, the new passes were higher. So too, considerably, was the intervening terrain. But it was this that made the passes less formidable. The mountains were not being circumvented but the deep, glacier-choked bends of the Shyok were.

Cayley reported enthusiastically on this new route. It had another advantage in that its point of departure in the Changchenmo valley could be as easily reached from the British provinces of Kulu and Lahul as from Leh. In fact, trade from British India could flow through Kulu via the Changchenmo route to Yarkand, completely by-passing the Maharaja's customs officials in Leh. This idea was so attractive that, in 1870, a special treaty would be extracted from the Maharaja, which would elevate this route to the status of a 'free highway', to be dotted with supply depots and rest houses and to be jointly supervised by a British and a Kashmir official.

In 1868 Shaw, as he thankfully drew ahead of Hayward and felt his way forward from the Changchenmo valley, had only Cayley's first year's reconnaissance to go by. This indicated that his next chance of obtaining food and grazing and of seeing another human being lay on the Karakash river, two hundred miles to the north. In between lay the dreaded Lingzi-Thang and Aksai Chin. These plateaux with their resounding names are usually described as deserts or plains. Photographs of them show an utterly featureless expanse of pebbles as flat as a dead calm sea. A group of tents and dejected horses adds no scale. They stand without context or location; they could as well be ten miles further back. In the hazeless atmosphere there is no perspective. Seemingly beyond the horizon, like faint relics of an earlier exposure on the same frame, lie low, unconvincing hills. They serve only to emphasise the dreary flatness of the scene. Yet this is still a part of the Western Himalayas. The hills are spurs of the Kun Lun and Karakorams, and the plains themselves are as high as the cruising altitude of a short-haul jet. To this day, only a handful of Europeans have ever crossed these stony wastes. Shaw, sadly, describes them without feeling. It was too cold to write, too cold even for his frozen fingers to pull the trigger on the rare antelope or wild yak. Fuel was so scarce that there was not enough to thaw snow for the thirsty ponies. And for days on end there was no grazing. Even his yaks began to collapse with exhaustion.

It is one of the mysteries of exploration in the region how anyone can have enthusiastically recommended such a route for trading purposes. Between the Changchenmo and the Karakash river the average altitude is 17,000 feet, the cold wind indescribable and the traveller's requisites of fuel, water, shelter and grazing, non-existent. Two years later Forsyth recognised it for what it was, a death-trap. He championed the Changchenmo route only as long as he had no experience of it. But those like Shaw, Hayward, Cayley and Johnson who actually travelled it, had no such excuse. One can see their indifference to the difficulties as a measure of their personal stamina, of their enthusiasm for a direct approach from India to Eastern Turkestan and of their assessment of the dangers of the classic route. But still one would have thought that its very dreariness would have depressed their ardour.

On November 7th, ten weeks after leaving the last inhabited spot in Ladakh, Shaw scrambled over a spur in the Karakash valley and spied an encampment of Kirghiz shepherds. Smoke rose from their black dome-shaped tents and a man in a long tunic and boots was tending his yaks and horses. 'I can't describe', he wrote, 'my sensation at beholding this novel scene. Now at length my dreams of Toorks and Kirghiz were realised and I was coming into contact with tribes and nations hitherto entirely cut off from intercourse with Europeans.'

This is typical Shaw. In education he was a cut above the average tea planter, but he always wrote in a consciously popular style—one which was later adopted by his nephew, Francis Younghusband. His idea was to attract the widest possible interest in his travels and so rally support for the notion of trade with Eastern Turkestan. In the process geography suffered. He described the Indus as rising in Lake Manasarowar, an idea to which Moorcroft had put paid half a century earlier. He managed to confuse Wood's Lake Sir-i-kol with the Sarikol region just west of Yarkand, and he greatly over-simplified the whole mountain structure by referring to it as just one range.

A more pardonable aspect of his treatment was the attempt to romanticise, first about the peoples of Eastern Turkestan and then about the place. His first meeting with Yarkandi traders had taken place in Leh. Beside the cringing Indians and the buffooning Ladakh-is, he had immediately recognised in the 'Toorks' men like himself. They were tall and dignified and fair as Englishmen. They looked you straight in the face, relished a hearty joke and were above all 'good fellows'. These notions were confirmed when he at last reached Shahidulla; the fort, at the junction of the Karakash river with the main Karakoram route, had been abandoned by the Maharaja in the previous year—he had no more right to it than I do, commented Shaw—and reoccupied by the Yarkandis as their southernmost outpost. They were expecting Shaw and he was well received. He was housed in the best room and given a foretaste of the hospitality in store. The only problem was conversation. He spoke a little Persian, but even that language, the lingua franca of educated Asiatics, was rarely heard in Eastern Turkestan. Instead, communication was at first carried on by signs, amidst a lot of back-slapping and digging in the ribs. 'They are just like public schoolboys, of boisterous spirits but perfectly well-bred.' And Shaw was flattered that they continued to distinguish between him—'they call me a good fellow'—and his Indian servants whom 'they treated as animals of some sort, monkeys for instance'.

However, beneath all the camaraderie, two disquieting suspicions arose in his mind. One was that he was their prisoner. No suggestion of this was permitted to mar good relations. Shaw was anxious not to put matters to the test and so were the Yarkandis who assured him that their close attentions were simply because of the honour in which he was held. Naturally he was under observation; everything depended on making a good impression and not arousing their understandable suspicions. But then why, if he was such a distinguished visitor, was he being detained so long at Shahidulla? Daily messengers streamed back and forth across the Kun Lun to Yarkand but still there was no sign of his being allowed to proceed.

And every day mattered. 'The thorn in my flesh', 'my chief source of anxiety' and 'the incubus that constantly weighs upon me' was Hayward and his imminent arrival. Ten days they had agreed on and, sure enough, on November 19th Hayward's approach was reported. Shaw had managed to hint to his hosts that the arrival of another Englishman was possible. But he had not felt his own position strong enough to urge Hayward's case. On the contrary, he had mentioned him only to safeguard himself and then in terms that were apologetic if not defamatory. In one account he claims that Hayward's arrival upset matters just at the point when he was about to be allowed to proceed. This is demonstrably untrue. A communication from Yarkand refusing him entry had arrived a day or two before Hayward.

His plea against this order, which might have been influenced by Hayward's appearance, was actually successful.

It may be that Hayward too broke his side of that agreement made over dinner in the Changchenmo. For on arrival, he threw Shaw into a rage by announcing that he was in Shaw's employ. This seems out of character on the part of the unbending Hawyard but, if true, it would certainly have jeopardised Shaw's chances. According to the latter, the damaging statement was only suppressed by a lucky coincidence; a chain of interpreters had now been set up—it went from English to Hindustani to Tibetan and so to the Turki of the Yarkand-is—and it started with one of Shaw's servants who, in his master's interests, tactfully omitted it.

Hayward, as if in quarantine, was housed some distance from the fort. Shaw had the satisfaction of seeing most of the guard moved over there. Puzzled as to how things stood, Hayward repeatedly tried to arrange a meeting, but Shaw refused lest he compromise himself with his new friends. They could communicate secretly through their servants but the only advice Hayward got was to go away. Shaw's position was doubtful; Hayward's, he claimed, hopeless.

And so it must have seemed when a few days later Shaw, dressed in Turki style with a turban, cloak and high boots, was seen to ride out north towards Yarkand. Hayward remained stuck in Shahidulla. In despair he thought of returning to Leh but the passes thence to Kulu or Kashmir would be closed by the time he reached them and he had no wish for a dreary winter in Ladakh. Nor of giving Shaw the satisfaction of knowing he had taken his advice. Hayward was a man of controversial character, but all were agreed on his inexhaustible energy. Anything like confinement was anathema to him and, typically, the day after Shaw rode out of Shahidulla, he did too. He gave the slip to his guards and made off west into the mountains on an excursion of which, as a feat of exploration and endeavour, it would be hard to find the equal.

In regard to Eastern Turkestan, Hayward's motivation was altogether different from Shaw's. As a young subaltern lately of Her Majesty's 79th Regiment,* the Cameron Highlanders, he had presented himself to Sir Henry Rawlinson in London as being 'desirous of active employment . . . on any exploratory expedition'. The doyen of Asian geographers and Vice-President of the Royal Geographical

* Not, apparently, the 72nd Regiment, as Rawlinson would have it.

Society suggested the cities of Eastern Turkestan and the Pamirs, and he drew up a memorandum for Hayward's guidance. Unfortunately this is lost, but whatever it said it is clear that Hayward interpreted his mission as primarily one to the Pamirs. He was intrigued by the question, reopened more than once since Wood's day, of the source of the Oxus and of what he calls 'the lake system of the region'. No one had actually repeated Wood's feat but there were native accounts of other important Oxus tributaries and other lakes on the Pamirs. The most notable of these was the vast Lake Karakul. Hayward saw himself emulating the discoveries of Speke and Baker, and was confident that the true source of the Oxus lay in this lake. But Yarkand and Kashgar never feature more prominently in his plans than as possible springboards from which 'to bag the Bam-i-Dunya'.

The whole subject of Hayward's emergence as an explorer is puzzling. He had some experience of travel, he was a fair draughtsman and he could use survey instruments. But beyond that we hear of no other qualifications. He was anxious to set off for India immediately, yet his funds were negligible till the Royal Geographical Society sponsored him with £300. Given that in the heyday of exploration they were more casual about these things, it still seems highly improbable that an unknown adventurer could land a prize assignment as simply as Rawlinson makes out. Much more needs to be known of Hayward's previous career before the question can be answered. For instance, his experience of travel seems to have included other forays into the Himalayas. He had probably been to Baltistan and Kashmir.* He may well have already met and learnt much from Gardiner, and he certainly knew Peshawar and the foothills below Chitral. It was for Peshawar that he had made as soon as he reached India, convinced that the best route through the Himalayas lay from there up the Chitral valley to the Oxus, the Pamirs and Turkestan. This was not Rawlinson's idea; he was more excited about Johnson's route from the Changchenmo which, at the time,

---

* Rawlinson refers to Hayward's having crossed the Hindu Kush on sporting excursions. This is highly unlikely given the attitude of the Afghans in the 1850s and '60s. Probably Rawlinson meant the Great Himalaya north of Kashmir and not the Hindu Kush as we now understand it—the nomenclature of the various mountain systems was still very confused. This would have taken Hayward to Baltistan (Skardu), already a popular region with sportsmen.

he mistakenly imagined to be the 'royal road' via Rudok which Moorcroft had reported. The Chitral route was much closer to the thinking of Burnes and Wood and, above all, Gardiner.

So too, were Hayward's notions of disguise. He was well aware, without the reminders from Shaw, that his credentials as a traveller were likely to arouse intense suspicion in an Asiatic mind. He was not, like Shaw, a merchant nor had he any political or religious standing. Yet traders, envoys and pilgrims were the only recognised classes of traveller. Rather than disguise himself as any of these, he chose to cut the more dashing figure of a Pathan (i.e. Afghan) mercenary— exactly what Gardiner had been during his greatest journey. It is perhaps also worth a note that Gardiner's famous outfit of tartan from turban to toe was said to originate from the Quartermaster's stores in the 79th Regiment, the Cameron Highlanders. That was Hayward's old regiment. Had they exchanged clothes as well as ideas, or is this simply a coincidence?

Hayward is the only professional explorer to figure in this account. As distinct from the surveyors, the naturalists, the political agents and the men of trade, the classic explorer had a simple brief, to make a name for himself. True, he hoped to further geographical science. This was essential, but it was the means rather than the end. The successful explorer was now revered as a popular hero, in a way that would have seemed as incomprehensible in Moorcroft's day as it does today. For the most part they were difficult men. They exhibited something of the arrogance and showmanship of a screen hero. Look at the photograph of Hayward. Bearded, strung about with sword and shield and resting on a spear as if it were a shepherd's crook, he looks like some wild evangelist. Burnes was always rather self-conscious about his portrait in Bokharan costume. Vigne donned his Balti outfit for the artist with the jaunty glee of an amateur. But Hayward, brooding and indomitable, looks deadly serious.

The outfit is not his Pathan disguise but, judging by the weaponry and the corkscrew markhor horns, that of Yasin where he was to meet his tragic end. The picture was probably taken by George Henderson who in 1870 would accompany Forsyth to Yarkand. He is one of the few who got to know Hayward. They met in Srinagar and spent several nights practising their astronomical techniques together. Hayward struck Henderson as the most dedicated and indefatigable of explorers. His passion for the unknown was infectious and his impetuosity such as almost to invite danger. To give Shaw his due, it can have been no fun having such a man on his tail. When Rawlinson had decided that 'if any Englishman can reach the Pamir Mr. Hayward is the man', he was referring to his protégé's dedication. Others called it madness. Hayward flirted with danger as determinedly as most dodge it. To any companion he was a standing liability. There was a chilly aura of impending disaster about him; it caused other men to keep their distance. We know of no friends or colleagues, and, as one peers into that cluttered photograph at the bony brow knitted into a frown between deep-set eyes, it is not hard to understand why. He looks a most unsettling companion.

Given the choice of two routes, one more dangerous than the other, he naturally chose the former. So it was with the Chitral route to the Pamirs. In terms of the passes involved he and Gardiner were eventually proved right; it was the lowest of all the routes through the mountains. But, in the 1860s, it was also the most lawless. Only the direct intervention of the Lieutenant-Governor of the Punjab had prevented him from following it. Thus, as a second and distinctly soft alternative, he had turned to the Leh–Yarkand route, hoping to penetrate the Pamirs from Eastern Turkestan.

Travelling much faster than Shaw's heavily laden caravan, he had scouted a new route between the Changchenmo and Karakash rivers, and sorted out the course of the latter which Johnson had mysteriously confused with the Yarkand river. This also drained from the Karakorams through the Kun Lun into Eastern Turkestan, and it was by way of completing his survey of the hydrography of the region that he broke out of confinement in Shahidulla and set off west to find the upper reaches of this river.

He was gone for twenty days. In that time he covered on foot three hundred miles over the roughest terrain imaginable. In one continuous march of thirty-six hours he did fifty-five miles. And all this at considerable altitudes, climbing and descending spurs of two of the greatest mountain systems, fording icy rivers and in the below freezing temperatures of early December. Hunted by the soldiers who had been sent after him from Shahidulla and by the robbers who infested the region, he had to douse his camp fire as soon as a meal was cooked. Yet he was still without a tent, and at the source of the Yarkand river he recorded fifty degrees below freezing. Food sufficient for only a week had been taken. He and his porters eked it out till it was finally exhausted. This left just the yak that had carried it. There was no fuel for a fire, but the beast was killed and the famished Ladakhis fell to on the raw carcase. Hayward hesitated; later he admitted that he, too, had joined in.

Eventually, striking the main Karakoram route, he restocked from a passing caravan, met up with the soldiers from Shahidulla and, on learning from them that he had been given permission to proceed to Yarkand, made a final dash over the Suget pass, thirty-three miles in a day, back to starting point.

As a feat of endurance the journey speaks for itself. To Hayward it was a necessary baptism. He went on to Yarkand fully confident that nothing could stop him reaching the Pamirs. He had also made a considerable contribution to geography. He had discovered a new pass over the Kun Lun which, together with his new route from the Changchenmo, made a complete and, he thought, far easier crossing of the mountains. He had found the source and explored the upper reaches of the Yarkand river and he had followed it to where the route from the Mustagh pass, which Godwin-Austen had attempted, joined its banks. Until Younghusband's journey of twenty years later, Hayward was the only authority on the wild gorges and desolate valleys between the Karakorams and the Kun Lun. It was for this excursion, as much as for reaching Eastern Turkestan, that the Royal Geographical Society awarded him a Gold Medal—an indication that people were at last beginning to realise that the mountain complex was, itself, a formidable arena for exploration and not simply a barrier to the unknown lands beyond.

Meanwhile Shaw, happily rid of the thorn in his flesh, had entered Yarkand. Like all the others who had panted to the crests of the ranges between India and Central Asia, he had breasted the Sanju pass in a fever of excitement. Like them, he had peered hopefully at the northern horizon and like them he had seen nothing in particular. But he was not disappointed.

> The first sight was a chaos of lower mountains while far away to the north the eye at last rested on what it sought, a level horizon indistinctly bounding what looked like a distant sea. This was the plain of Eastern Turkestan and that blue haze concealed cities and provinces which, first of my countrymen, I was about to visit.

The first glimpse of Yarkand was less exciting. A long blank mud wall stretched across the flat desert. Pagoda-like structures topped the corners and main gateway and a few leafless poplars rose above it, leaning with the prevailing wind. The only other feature was a gallows-like structure which was as high again as the walls. From a distance it was hard to appreciate the scale, but the walls were in fact forty feet high. A moat encircled them and a good size road was carried round on top. As for the scaffold-like structure, it was precisely that. Hangings were high and frequent in Yarkand.

But more than this was needed to still Shaw's excitement. He had already seen something of the land and its people, and he was convinced that it really was an Eldorado. Explorers are prone to describe their discoveries in glowing terms. As soon as they tread new ground the rose-tinted spectacles are on. The grass looks greener, the air smells sweeter and the people are more romantic. It must also be remembered that for the last three months Shaw had been in the mountains. A cultivated field, a row of shops, a warm hearth and a square meal were all joyous novelties. Added to this, he was being welcomed in no meagre style. 'A swell moghul' accompanied him over the Kun Lun. Even his servants were mounted at state expense and, on arrival at the first village, he was saluted with guns and regaled with an address of welcome from the king. The road had been specially repaired. In the villages it was lined with people and in the countryside lavish picnics awaited him at every stage. These *dasturkhans* were a distinctive feature of Turkestani hospitality. They varied from a humble spread of dried fruits and tea to a veritable cornucopia of local produce. Hayward's first comprised 'two sheep, a dozen fowls, several dozens of eggs, large dishes of grapes, pears, apples, pomegranates, raisins, almonds, melons, several pounds of dried apricots, tea, sugar, sweetmeats, basins of stewed fruit, cream, milk, bread, cakes etc., in abundance'. Shaw mentions that, by the time he reached Yarkand, he had a whole flock of uneaten sheep, his wardrobe was stuffed with gaudy robes of honour—one was given at each meeting with an official—and he had not been allowed to pay for a single thing.

Under the circumstances he looked on his surroundings with a kindly eye and was duly amazed by the fertility of the country and the prosperity of its people. It was like an Asiatic Holland, less bare than some of the French provinces and with villages that reminded him of home. In the farmyards the cocks crowed and the ducks quacked. Orchards gave way to well tended fields and ditches gushed with water beneath fine trees. The country folk were rosy cheeked and cheerful. Market day brought them thronging to the nearest village and nearly everyone rode a horse or a donkey. Compared to

India there was no poverty, no beggars and no squalor. The people 'looked respectable, brisk and intelligent'. In the cities the bazaars were orderly and well stocked. There were bakers, butchers, saddlers, goldsmiths, tailors and even restaurants. Everything, except tea, was reasonably priced and the whole population showed surprising sophistication in their shopping. Here, in short, was no backwoods country awaiting the prospector and the trapper, nor a mouldering civilisation ripe for the archaeologist, but a modern and thriving state, able to supply most of its basic needs yet commercially experienced and traditionally outward looking. It felt a bit like Europe and, as a potential market, it was of European significance.

Now this was all very well. Eastern Turkestan is certainly different from India. It belongs to the tough and restless world of the great continent and not to the gentle, fatalistic ambience of the subcontinent. The horse, not the cow or bullock, is the basis of the traditional economy. Horseflesh is a prized delicacy, horse-drawn carts ply rapidly between cities, and both men and women live all the year round in high riding boots. It gives the whole place a different feel, a sense of bracing activity, and to the people, an air of rugged independence. India lies to the south, but to the north lies Siberia. Tolstoy* was attracted to Eastern Turkestan but Tagore would have hated it.

It is not, however, an Asiatic Holland. Shaw, and to a lesser extent Hayward, was blinded by his own enthusiasm. The local name for the country was then Altyn Shahr, the Land of the Six Cities. This is a good description. Six cities, oases in fact, dotted round the Takla Makan, a great western extension of the Gobi desert. Shaw mentions sandy, uncultivated tracts as if they were narrow bands breaking the monotony of the endless villages and hamlets that he so minutely describes. It is, of course, the other way round. Nine-tenths of the country are uncultivable and the remaining tenth is at the mercy of the shifting desert sands which have buried whole cities in the past.

*   *   *

Shaw had left Shahidulla happy in the knowledge that he had seen the last of Hayward. In a letter smuggled back to Leh in a sack of flour, he told his sister 'he has not prepared his way as I have' and 'will not be allowed to come on'. He also said that in Shahidulla they

* Tolstoy told Henry Lansdell in 1887 of his interest in Eastern Turkestan, and would have liked to accompany him there.

had been prevented from meeting. This was certainly Hayward's impression, but Shaw later admitted that it was he who had rejected the idea of a meeting. As usual he feared it might compromise him.

Arrived in Yarkand, he learnt that Hayward's case was not hopeless. The latter's letter seeking permission to enter the country had been written in English and was still unread. Shaw was asked to translate and noted with surprise that Hayward now claimed to have travelled the eight thousand miles from England for the purpose of trade. Where then was his caravan of merchandise? The Yarkandis were not going to believe this story for a minute. And when news came of Hayward's escape from Shahidulla, he was confident that there was an end to the matter.

He didn't realise that permission for Hayward's onward journey had now been given; he was, in fact, just starting across the Kun Lun. In due course he reached Yarkand, and for a week the two men were lodged within a hundred yards of one another. But no communication took place. In his narrative Shaw only acknowledges his rival's presence in order to take a swipe at him. Hayward had made a fool of himself by sitting before the governor with his legs stretched out 'in defiance of all oriental etiquette'. On January 4th (1869) Shaw was escorted on to the capital, Kashgar. Two months elapsed before Hayward joined him.

With both men safely into Turkestan, their main concern shifted to whether they would ever be allowed to leave. They both knew of Adolph Schlagintweit's fate, and they soon began to realise that entering the country was less than half the battle. To earlier travellers the Chinese policy that rigorously excluded all visitors had seemed unreasonable. But Moorcroft, for one, must have appreciated that it was a good deal more humane than the deceptive overtures of a Murad Beg. The time-honoured custom of the Mohammedan cities of Central Asia was to welcome all comers indiscriminately, but to show great reluctance in allowing them to leave. Not only Hindu traders and Pathan mercenaries had been ensnared in this way. A few Englishmen, Stoddart and Connolly in Bukhara and Wyburd probably in Khiva, had blundered into the trap and never been seen again.

Eastern Turkestan, now under Mohammedan rule, was every bit as dangerous a place. The exclusiveness of Pekin's rule had given way to a reign of terror and treachery which had made the land a byword for anarchy and mindless slaughter. In Kashgar, the besieged Chinese were said to have eaten the rats and the cats, then their own wives and children, before finally surrendering. In Yarkand, the Chinese governor had assembled his family and officials for a final feast, listened to their suggestions for the terms of surrender and then quietly tapped out his pipe on to a prelaid trail of gunpowder that blew the gathering, the fort and Chinese rule to kingdom come. This ushered in a decade of confused power struggles in which, even by the standards of Central Asia, life was cheap in the Land of the Six Cities. Men were of no more account than sheep—and considerably less than horses. 'Emphatically', wrote Shaw, 'every man carried his life in his hands.'

The successful contender who by 1868 was undisputed ruler of an almost reunited land was an outsider, Mohamed Yakub Beg. It was he who commanded the army from Khokand that Johnson had reported outside the walls of Yarkand in 1865. He was an invader rather than a patriot and his rule relied heavily on the Khokandi merchants and soldiers who had always figured prominently in Eastern Turkestan. But regardless of his origins and the dubious means by which he had risen to power, Yakub Beg was soon to be regarded by the British as the hero of the hour. His achievements would be compared to those of Ranjit Singh and he would be hailed as the greatest conqueror in Central Asia since Timur and Baber. For a time there was even talk of his uniting the Islamic states of Central Asia, stemming the tide of Russian advances and extending his rule as far east as the Great Wall. His ambassadors were welcomed in Calcutta, St. Petersberg, Istanbul and London.

Yet by the end of the decade he was dead, his rule was discredited, his achievements dismissed, and the Chinese were back in the cities of Eastern Turkestan for good. Far from moulding history he was now seen as a freak of circumstance, a clumsy opportunist who overreached himself. His subjects were said to be pleased to see the alien Chinese back again and neither the British not the Russians greatly regretted his departure.

The attitudes of Shaw and Hayward towards their host anticipate this ambivalent treatment. Shaw had his first audience soon after he arrived in Kashgar. He found the king 'friendly and courteous', 'fatherly and affectionate' and 'an awfully plucky leader . . . beloved by his people'. Under his firm but wise rule the country was enjoying unheard of prosperity, the roads were as safe as in England and new public works were being undertaken. Hayward went along with this, calling him 'a man of extraordinary energy, sagacity and ability' who had already accomplished wonders. But at the same time, though enjoying lavish hospitality at the king's expense, both men had their reservations. In Kashgar, Shaw's suspicions that he was more a pampered prisoner than a respected guest were amply confirmed. Except for the occasional official visit, he was not allowed out of his quarters for three months. Night and day he was under close surveillance. His visitors were vetted and his only acquaintances were the guards and spies who attended him. If his observations on the country and its people were wide of the mark this was partly because he saw so little of them; nothing, in fact, beyond the high road to Yarkand and thence to Kashgar. If in vain one looks for news of the spirited ladies of the land, about whom Burnes had heard so much, it is because he was never allowed near the bazaars of the two main cities. For his assessment of Yakub Beg he relied heavily on information gleaned from men who owed everything to the new ruler and were mostly fellow countrymen from Khokand. His descriptions of the tall dignified 'Toorks' are not descriptions of the natives at all but of their Khokandi rulers. He had virtually no contact with the local population.

Occasionally rumours of the true state of things did reach him. His Indian and Ladakhi servants were allowed far greater freedom. They wandered into the bazaars and brought back stories of a very different kind. Yakub Beg's rule was based on a literal interpretation of Islamic law. Anyone not punctuating his day with the proscribed breaks for prayer or dodging attendance at the mosques, any woman going unveiled or man unturbanned, was summarily flogged. For a minor offence mutilation was common. For a second there were the gallows, as prominent in Kashgar as in Yarkand. The 'fatherly ruler' with his austere habits was given to fits of uncontrollable rage in which, with his own hands, he had been known to strangle a subject. One of his greatest achievements—Shaw and others honestly admired it—was the perfecting of a system of secret police. Informants were everywhere and any political divergence meant certain death. Lansdell, the man who suffered so much from the cold on the Suget pass, found in Eastern Turkestan an undreamt of refinement in the field of punishment and torture. He also claimed that the king's orthodoxy was a sham. He relished the suppression of prostitution, destroying whole streets of brothels and murdering their inmates. Yet his private life was not edifying. He had a harem of

21 Robert Shaw, Dr. George Henderson and Sir T. Douglas Forsyth in Yarkand

22 Andrew Dalgleish and his Yarkand household including Abyssinian servant and Yarkandi wife

23 A caravan in the Karakorams

24 The Indus Valley near Leh

three hundred, and two of its youngest inmates, mere children, died as a result of his brutalities.

During his long detention Shaw began to give ear to such rumours. It would have been suicide to pursue them but, as the dreary days wore on full of uncertainty and gloomy forebodings, he at last conceded that 'darkness is the rule of the land'. He played along with the boisterous humour of his guards but vented his frustrations on a home-made set of dumbells and confided his worries only to the cats of Kashgar. They seemed to recognise a fellow outcast.

Hayward, still in Yarkand but under the same surveillance, was less resigned. To him, forced detention was like a lingering death. He was still unable to provide a convincing explanation for his presence in the country, and his hopes of reaching the Pamirs were fading fast. Unlike Shaw he found it hard to hide his true feelings. In February he decided once again to put matters to the test and broke out of captivity. According to rumours reaching Shaw, he drew his revolver and set off on a promenade of the bazaars. But troops were immediately deployed round the city and Hayward meekly returned to his lodging. He accepted that escape was impossible and made no further attempts. Judge of his sentiments from a caption on the back of one of his watercolours.

> House in the fort, Yarkand.   22nd. Feb. 1869
> The house in which they entertain their friends so hospitably in the fort of Yarkand. First enticing them into the country as guests and then confining them like prisoners. Not to destroy the peaceful harmony of the scene, a guard of eight sepoys and a Panjabashi [officer] have been judiciously omitted. This is the way they treat their guests in Turkestan.

At the end of February Hayward was summoned to Kashgar, and at about the same time another visitor from India also arrived there. This man, known as Mirza Shuja, was a native said to be in British employ. Neither Shaw nor Hayward had ever heard of him, but he soon made contact with the former, writing in English to ask if he could borrow a watch. He also asked what day of the month it was; this and the watch were vital to anyone trying to establish his exact position from the stars. Characteristically, Shaw was highly suspicious. The man must be 'a dangerous imposter' and the whole thing some deep-seated plot to implicate him as a spy. His answer was a stony silence; not even the date would he reveal. Hayward might have been a better bet but, perhaps discouraged by Shaw's cold-shouldering, the Mirza does not appear to have approached him.

The whole affair must indeed have been a bitter blow to the Indian. He was, in fact, one of a select group of native surveyors trained by Montgomerie of the GTS to carry out route surveys in the unknown beyond India. These intrepid travellers made their greatest mark in Tibet, but under code names like The Havildar and The Mullah a few operated in the Mohammedan countries further west. The Mirza had come to Kashgar from Kabul following Wood's route up the Oxus and continuing across the Pamirs. He questioned Wood's choice of the Sir-i-kol branch instead of the Sarhad as the source of the river, and furnished the first narrative of a crossing of the Pamirs since Benedict de Goes. All this had taken him well over a year, and it must have been with great excitement that, on arrival in Kashgar, he heard of the presence there of two Englishmen. It was reasonable to expect that at the very least they would smooth his relations with the Kashgar authorities.

March 1869 was, for all three men, the blackest of months. Shaw's promised second meeting with Yakub Beg had still not materialised. He was watched more closely than ever and the air was thick with disturbing rumours. The Mirza was said to be chained to a block of wood like a common criminal. Hayward had had his first audience with Yakub Beg and had been led to understand that there was absolutely no hope of being allowed on to the Pamirs. From the roof of his house he could see the range that bounded them on the east. It was only sixty miles away. To someone who had travelled half way round the world to get there, and whose 'mouth watered at the very name of the Bam-i-Dunya', it was a bitter disappointment. He gave vent to it by telling the king's representative exactly what he thought of their hospitality. At the same time he wrote to Shaw, the first letter since Shahidulla.

Now, if ever, was the moment for the two men to draw together. The cards were on the table. Shaw had no more hope of winning commercial concessions than Hayward did of being allowed to tackle the Pamirs. Their one and only objective was the same, to get out of the country alive. They were also agreed about the duplicity of their captor. Shaw makes what for him, after his extravagant praise of the king, must have been a difficult admission; that Yakub Beg was forcibly detaining them for his own purposes. As usual Hayward was more outspoken. Yakub Beg was 'the greatest rascal in Central Asia'. At their meeting the king had ominously referred to the fate of Stoddart and Connolly in Bokhara—they were confined in a pit full of snakes and scorpions and then executed—and to that of Adolph Schlagintweit who was knifed to death just a mile or two from the gates of Kashgar. Hayward believed that his life and Shaw's were also at stake and he told Shaw so in his letter.

Yakub Beg's treatment of his two captives was certainly alarming. One minute they were being lulled by promises and lavish *dastur-khans*, the next they were being neglected, reprimanded and threatened. And all the time they were a prey to the fears, fed on rumour, which mushroom in the minds of men kept in solitude. On Shaw at least this treatment worked. After three months he was feeding out of the king's hand like a turkey in early December. He had always been the more pliant and obsequious; he justified it by reasoning that therein lay his one chance of safety. Hayward, more alarmist and less receptive to the soft glove, was unimpressed. He recognised that he had no chance of allaying the king's suspicions about why he had ever come to Eastern Turkestan, and increasingly he stood on his dignity as a disinterested and independent British traveller. To Shaw's way of thinking this made him more of a liability than ever. There was no question of their pooling their efforts and the first letter went unanswered.

A second letter reached Shaw via their servants on March 24th. In it Hayward at last gave Shaw a piece of his mind. He wrote from the heart, a bitter and facetious outpouring unlike anything else in his extant correspondence. Shaw, hoping to show the sort of man he had to deal with, included an extract in his book. It is all that survives, but what an eloquent, prophetic and tragic passage.

And now I'll sketch your future for you [wrote Hayward]. You will return to be feasted and fêted, as a lion fresh from Central Asia. You will be employed on a political mission to Eastern Turkestan; you will open out my new trade route with countless caravans; you will become the great 'Soudagar' [merchant] of the age, and drink innumerable bottles of champagne in your bungalow on those charming Lingzi Thang Plains; you will write endless articles for the 'Saturday' and a work on the geology and hydrography of the Pamir plateau; you will win three Victoria Crosses and several K.C.B.s and live happily ever after. In contradistinction to all this, I shall wander about the wilds of Central Asia, still possessed with an insane desire to try the effects of cold steel across my throat; shoot numerous *ovis poli* [Marco Polo sheep] on the Pamir, swim round the Karakul Lake, and finally be sold into slavery by the Moolk-i-Aman or Khan of Chitral.

'These predictions are very singular', thought Shaw in 1870. He could have been referring to the predictions for his own future. He did return safely, he was lionised in England and his next assignment was on a political mission to Eastern Turkestan. But, in 1870, it was Hayward's prediction of his own future which struck Shaw as curious. For already Hayward had tried the effects of cold steel across his throat. His mangled corpse lay under a pile of stones beneath the glaciers of Darkot in the Hindu Kush. The supposed instigator was the 'Moolk-i-Aman or Khan of Chitral'.

13. A Bad Beginning

Reading the accounts left by Shaw and Hayward of their captivity in Kashgar, it is easy to overlook the seriousness of their plight. Their outbursts of indignation and alarm are so sandwiched between long eulogies on the country and its people that they seem out of place and unjustified. This is because both men wrote in the light of subsequent events. What, at the time, was a horrifying reality was soon remembered as no more relevant than an uncomfortable nightmare. For in April 1869 Yakub Beg suddenly changed his tune. Rumours of their impending release were nothing new to the two prisoners, but almost before they dared credit a new wave of these, they were on their way back to Yarkand. The king was not simply washing his hands of them but evincing signs of what Shaw took for genuine affection. There were apologies for the long delay, promises of commercial co-operation and fond farewells. It was even suggested that an envoy from Kashgar might accompany them to India. In Yarkand they waited for the opening of the passes back to Leh, and left on May 30th. At about the same time the Mirza was released in Kashgar.

It could be that Shaw's compliance, or Hayward's obduracy, had somehow paid off. More plausibly, Yakub Beg's change of heart had to do with *realpolitik*. The departure of the Chinese had made Eastern Turkestan, like the Khanates of Western Turkestan, fair game in the gradual Russian absorption of Central Asia. And ironically, it was to the Russians that Shaw and Hayward owed their safety.

The Great Game is usually divided into two distinct phases, the first being more or less coterminous with the mercurial career of Alexander Burnes, and the second beginning quarter of a century later in the late 1860s. After the disastrous British intervention in Afghanistan, and a simultaneous and equally abortive Russian expedition against Khiva, the Tsar's designs on Central Asia were for a time viewed by the British in India with more complacency. Men like Burnes and Moorcroft were written off as dangerous alarmists so long as Russian expansion was seen to be directed not towards Kabul but Constantinople. The Crimean War of 1854–56, far from intensifying Anglo-Russian rivalry in Turkestan, seemed to prove the point. Turkey and perhaps Persia were thought to be the Russian priorities, but not India. This was quite correct, but it did not mean that St. Petersberg had forgotten about Central Asia. On the contrary. While the Indian government was busy with the annexation of Sind and the Punjab and the implications of Kashmir's feudatory status, the Russians were pushing quietly south to the Aral Sea and then up the Jaxartes towards Khokand. Simultaneously, another line of advance was opened from Siberia south towards the Tian Shan. There was a slight lull during the Crimean War but the Indian Mutiny a year later renewed Russian interest in the subcontinent; from now on it would be a premise of all Russian thinking about India that the British Raj would be overthrown from within as soon as they were near enough to foment trouble and afford sanctuary. Under this new impetus, the advance south continued. In the early 1860s the gap between the two earlier lines of advance was filled in and a fort was erected on the Naryn river only four marches north of Kashgar. So much for the groundwork. The headline news, which heralded the second phase of the Great Game, soon followed. In 1865 they took Tashkent, in 1866 Khodzhent and in 1868 Samarkand fell. Of the three Central Asian Khanates, Bukhara and Khokand were now at the mercy of the Russians and an expedition was already being prepared to march again across the deserts to Khiva.

While the forebodings of Burnes and Moorcroft were in least favour, they were actually coming true. The British failure to take up the gauntlet of commercial competition and political influence in Central Asia had enabled the Russians, by the late 1860s, to wipe the board. Instead of two thousand miles separating the two empires, there were now barely five hundred. All that remained were the Pamirs, with Afghanistan and the Turkoman country to the west, and Eastern Turkestan to the east. From a Russian point of view, Afghanistan was too closely aligned with the British and the Pamirs were largely unknown. But Eastern Turkestan was both an obvious and attractive proposition.

No one appreciated this more than Yakub Beg. In 1853 he had himself conducted a valiant defence of one of the Khokandi fortresses against the Russians. He had then blazed a trail into Eastern Turkestan which the Russians might well follow. And in 1867 he had laboured with his own hands to throw up defences against them in the Tian Shan where there had already been frontier clashes. He had no illusions about the strength of the Russian forces and, though he prepared for defence, he also prayed for an accommodation. When, in 1868, a Russian envoy turned up in Kashgar he was therefore treated with respect. He left the city just before Shaw arrived, and he took with him to St. Petersberg a personal emissary from Yakub Beg.

According to reports gathered by Shaw, the Russians were seeking free passage for their troops through Eastern Turkestan, 'promising to turn neither right nor left but to go straight against their "brothers" the English'. Another, more realistic, version had it that they were after commercial privileges including their own premises and a permanent representative in Kashgar. Something similar had been promised in a Sino-Russian treaty of 1860 and in St. Petersberg it was hoped that the new ruler would be prepared to offer the same arrangement. Shaw says that Yakub Beg's reply was a categorical no but, more probably, he was willing to consider the idea if the Russians would guarantee his independence. Hence the despatch of the return envoy, and hence the detention of Shaw and Hayward while news was awaited from Russia.

Early in April 1869 the envoy returned. The Russians were not open to any such deal, and Yakub Beg turned with renewed interest to his two captives. At the time of their arrival it seems that the king held only the haziest notion of what lay to the south of his frontier. He was under the impression that the Maharaja of Kashmir was a great and independent sovereign. Indeed, from well-attested reports that every autumn he ejected the British from his domains, Yakub Beg was inclined to regard the Maharaja as the suzerain and the British as the feudatories. Shaw had eagerly set about disabusing him. He explained the British obsession with hill stations and summer holidays. He pointed out that no one in his right mind, the Maharaja included, spent the winter in the Kashmir valley, and he emphasised that the Maharaja was only suffered to rule for as long as his loyalty was unquestioned. Now, when Yakub Beg was at last prepared to listen, Shaw's propaganda was dramatically rammed home by the news that the Amir of Afghanistan, the most respected of Mohammedan princes, had just journeyed to India to seek arms and moral support. Under the circumstances, Yakub Beg decided that nothing could be lost by opening relations with such a bountiful power.

His envoy was not ready to accompany Shaw and Hayward from Yarkand but they left with the king's protestations of friendship for their *malika* (queen) and Lord Sahib (Viceroy) ringing in their ears. Hastily they revised their opinions of the 'greatest rascal in Central Asia' and congratulated themselves on having opened the way for further contacts. On the outskirts of Yarkand the two travellers came face to face for the first time since that dinner in the Changchenmo. Uneasily they travelled together as far as Shahidulla then, with no regrets at parting, made their separate ways to Leh.

Here more good news awaited them. Just before Yakub Beg had turned his attention towards India, the bureaucrats there had come to regard Eastern Turkestan in an altogether more favourable light. For in January 1869 the unadventurous Lawrence had been succeeded by Disraeli's unexpected nominee, Lord Mayo. A big, bold Irishman full of dash and charm, Mayo survived Gladstone's early return to power and brought to the question of British India's relations with her neighbours a mind wide open to new ideas.

Shaw, the planter, and Hayward, the explorer, had ostensibly no political interests. Lawrence would have put paid to their plans if there had been any likelihood of their engaging in diplomacy. Nevertheless, Hayward was the protégé of Rawlinson, the most vocal and best informed of the London watchdogs of Russian strategy, whilst Shaw was the friend and accomplice of Forsyth, than whom no one in India was more concerned about Central Asian affairs. Both travellers appreciated that where the explorer and the merchant might go, the envoy and the flag might follow. And both now made the most of such political and strategic information as they had gleaned. Official reports were submitted and Shaw soon had the ear of the new Viceroy.

As the Great Game entered its new phase, hegemony in Central Asia was no longer an issue; but the defence of India was. The precise nature of the threat was as ill-defined now as in the 1830s. The views of men like Rawlinson were ambivalent. In one breath they discounted any possibility of a full-scale invasion of India. Banish the thought; it was quite unworthy of any rational observer. Instead, what mattered was that Russia should be denied any further advances, and this was particularly important in the light the 1857 mutiny had shed on the loyalty of the Indian troops. But the old bogey of invasion was not so easily laughed away. In fact, it was more feasible now than ever. In the next breath Rawlinson was re-invoking it with lurid detail; '50,000 Persians, 20,000 Turkoman horse than whom there is no better irregular cavalry in the world', all officered by Russians and stiffened with formations of regular Cossacks, swarming into the Punjab. Apart from anything else this was much the best way of creating another wave of Russophobia which alone seemed to constitute a favourable condition for the formulation of a trans-frontier policy.

At the time—he had delivered this particular memorandum just as Shaw and Hayward were setting off for Yarkand—Rawlinson was more concerned about an attack from the direction of Persia. But he was also intrigued by the finger of Russian progress up the Jaxartes which seemed to be pointing ominously at Eastern Turkestan and Kashmir. The Royal Geographical Society was supposed to be strictly non-political, and it was they who sponsored Hayward. But Rawlinson, as his mentor, cannot have failed to rouse his interest in matters other than the purely scientific. Similarly, when in 1870, with Rawlinson about to be its President, the Society awarded Hayward its Gold Medal, it was not just a fair recognition of his achievements but also a good way of drawing further attention to his conclusions. For Hayward had gone still further than Shaw. He supported the reports of Russian interest in Eastern Turkestan and, from his greater knowledge of the mountains, insisted that a modern army complete with artillery could cross them; his new trade route was also a possible invasion route.

There were, then, good strategic reasons for reconsidering the role of the Karakorams as an impregnable barrier and for seeking an arrangement with Yakub Beg that would bring Eastern Turkestan into the pattern of India's defence. Mayo listened carefully to the reports of Shaw and Hayward. And he decided that Eastern Turkestan might play a role in the north equivalent to that of Afghanistan in the west, that is, be a friendly buffer state, kept strong and reasonably united on a planned diet of British support, both material and moral. To lull the suspicions of the Russians, and of the Liberals now in power at home—the party politics of Westminster were increasingly making themselves felt in India—his new policy was to be inaugurated by purely commercial overtures. This was where Forsyth and Shaw came in, for though they shared the strategic concerns of Rawlinson and Hayward, they also believed passionately in the value of the trade itself.

In late 1869 the envoy from Kashgar, who was to have accompanied

Shaw and Hayward, arrived in India. His main request was that an official representative of the government should return with him. Mayo leapt at the idea and Forsyth, assisted by Shaw and George Henderson, a doctor and naturalist in the tradition of Thomson, was deputed on the first ever official mission to Eastern Turkestan.* Shaw sped back from England where he had spent the winter enjoying the fruits of his labours and drumming up support for the opening of the Yarkand market. He hoped to join the rest of the party before they left Leh in July.

Meanwhile George Hayward too had been living out his 'singular prediction'. His goal was still the Pamirs, but denied both the Chitral and Yarkand routes he was now concentrating on yet a third, that via Gilgit and Yasin in the Hindu Kush. The Royal Geographical Society had put up another £300, and the government had paid £100 for his map and report on the journey to Eastern Turkestan. However they refused to get involved in his further plans. This time even he admitted that 'the danger is certainly great'; his new route combined all the tribal lawlessness of Chitral with the mountain difficulties of the Karakorams.

But danger was what Hayward relished. It was time to pull out all the stops. His energy during this last year of his life was prodigious. Invariably alone, pushing his luck further and further, he crashed about the mountains like a hounded stag. He had reached Leh in July 1869. In September he was writing from Murree (after a march of 300 miles), in November from Srinagar (150 miles), in January from Gilgit (150 miles), in March from Yasin (100 miles), in April from Murree (300 miles), in May from Srinagar (150 miles), in June from Gilgit (150 miles), and in July he was back again in Yasin (100 miles). It was as if he knew his days were numbered. And so, indeed, he did. He was safe from Yakub Beg, but the Maharaja of Kashmir had even better reasons for wanting him out of the way. Kashmiri agents, one of them a specially hired prostitute, had already made such determined attempts to poison him that he would accept food prepared by only one faithful servant.

All the explorers of the period, Cayley, Shaw, Forsyth and even Johnson, deeply mistrusted the Kashmir government. Kashmiri

* Mayo was at pains to emphasise that Forsyth's visit was neither official nor a mission. But since he was sent by the government specifically to discuss trade with Yakub Beg, one might ask what then the exercise was supposed to be.

traders were as bitterly opposed to British intervention in the Yarkand trade as they had been in Moorcroft's day. And the Maharaja was as sensitive about his frontiers as Vigne had found his father, Gulab Singh. But Hayward had still stronger reasons for concern. He gave it as his opinion that Russian spies were being entertained in Kashmir, and that the Maharaja had his own agents in Central Asia. Worse still, on the first visit to Yasin he had uncovered the grisly remains of a massacre of genocide proportions, by the Kashmir troops. He wrote a report on it in which, with his usual disregard for the consequences, he suggested that the British people should be made aware of what their loyal feudatory was up to. Someone took him at his word. An article, mentioning him as the authority, appeared in the press in May 1870.

At the time Hayward was in the Kashmir capital and about to set off again under the Maharaja's protection for the wildest corner of his state. It was rather like kicking the lion in the belly while he stuck his head in its mouth. His friends told him to return to India fast. Mayo sent a telegram to the Royal Geographical Society asking them to stop him from attempting the Pamirs again at all costs. He was a liability alive, but if he went ahead and got his throat cut there was no telling what the consequences might be. Hayward of course could not see it that way. He offered to sever all connections with the Society but he was not going to abandon his journey. He had put the Maharaja on the spot and believed the Kashmiris must now feel duty bound to see him safely on to the Pamirs. He left Srinagar in mid-June 1870.

A month later, on July 17th to be precise, another dinner party took place in the bleak Changchenmo valley. Hayward was absent; his route lay hundreds of miles to the west. But Shaw, along with his two companions, Forsyth and Henderson, played host. Their guests included Cayley, the Joint Commissioner for Trade from Leh, and their Yarkandi travelling companions, including the returning envóy and a cousin of Yakub Beg's. This time there was no shortage of food; the Forsyth mission was far better equipped than Shaw and Hayward had been. Given the season, the claret, too, should have flowed more easily. The only compromise that had to be made to their surroundings was that of sitting on the ground—there were not enough chairs to go round.

During the evening final arrangements were made for crossing the wilderness ahead. The baggage was reduced to a minimum. Seventy ponies were deemed unfit for the journey and some of their loads transferred to yaks. The remainder, most of it grain for the surviving ponies, was to be sent on by the Kashmiri governor of Ladakh. He was also in camp but, perhaps unwisely in view of what was to take place, he had not been invited to dinner. The evening went well. An alarm that the river was rising and flooding the camp came just after the meal was over and provided a convenient excuse for the party to break up. All retired early in preparation for a prompt start in the morning on the toughest leg of the journey.

'In contradistinction to all this', on the same night, three hundred miles to the north-west in a lonely spot at the foot of the Darkot glacier, George Hayward ate no dinner. Nor did he go to bed, but sat up through the long night writing by candlelight. In his other hand he held a pistol. On his collapsable table lay more loaded weapons.

At dawn he drank a cup of tea. All was quiet. He sniffed the morning air, then turned in for an hour or two's sleep. But the danger was not passed. He awoke to hard hands clutching at his throat. Bound and bullied he was led out on to the hillside. They took the ring from his finger, he muttered a prayer, a sword swept and his head fell from his body.

It all happened between the hours of eight and nine on the morning of July 18th. Well-breakfasted and with the self-conscious importance of men starting on a great endeavour, the members of the Forsyth mission were streaming out of camp. If Shaw looked back down the ugly Changchenmo valley it would not have been with nostalgia for his first visit. It is notable that his voice was never joined to the storm of protest and conjecture that arose over Hayward's murder. Forsyth and Cayley were loud in their accusations of the Maharaja. Others made out a strong case against the Chitral ruler, while the official line was that it was just a case of robbery with violence. Shaw kept silent. He had seen it all coming. If anyone was responsible, it was Hayward himself with that insane desire to try the effects of cold steel across his throat.*

News of Hayward's fate only reached the Forsyth mission two months later when they were on their way back from Yarkand. By then Forsyth had ample reason for holding the Kashmir government

* Hayward's last journeys and his murder belong to the story of the exploration of the Gilgit region. They are dealt with at greater length in a volume now in preparation by the same author.

responsible. From the day the mission left the Changchenmo things started to go wrong, and within a week they too were on the verge of disaster, all thanks to the intrigues of the Kashmir authorities.

It was almost axiomatic that any expedition heading north from Leh was swindled over the hire of ponies. The beasts actually inspected and hired were never the ones that finally materialised. There was always a shortfall and the condition of the animals, as much as the terrain, was responsible for the high death toll. Hence it was not surprising Forsyth found himself so short of serviceable ponies that he was dependent on the Ladakh government for forwarding the grain needed for those that were fit. The arrangement was that the governor would immediately send the grain on by yaks and would then wait in the Changchenmo until he was informed that the mission had safely reached the first Kirghiz habitations in the Karakash valley on the other side of the Aksai Chin. In the event he did neither of these things. Knowing that his arrangements were inadequate and that the mission would probably be brought to a standstill, he quietly withdrew to Leh. It was not, and never had been, in Kashmir's interest to foster relations between Calcutta and Kashgar.

By July 21st Forsyth's position was desperate. Even Shaw, who always contrived to see Yarkand and the route thence in the best possible light, admitted in his report for *The Times* that they were in 'a critical state'. The ponies were starving. There was no question of waiting for the yaks to come up since they had run out of fuel. And there was no word from the governor as to where the yaks might be. During the next three days over a hundred ponies died. They were so desperate that they tore at the baggage in search of grain. At night, driven wild by the cold and hunger, they careered through the camp and then quietly slunk into the tents to die beside, and sometimes on top of, their tormentors.

On the 23rd the mission ground to a complete standstill. Food for the men was also getting desperately short, but it was imperative that they halt while the remaining ponies cropped what they could. The opportunity was taken for a council of war in which it was decided that, in a last desperate effort, the party should be divided into three. All hopes of survival rested on the Yarkand envoy and his followers. They were given the best horses and told to ride post haste to Shahidulla and get help. To ensure their co-operation and assist their progress they were persuaded to leave behind their baggage, their four hundred muskets purchased in India and even their womenfolk. Forsyth with his companions and the Yarkandi ladies plodded on as best he could, and a third party was left behind on the Aksai Chin to guard the abandoned baggage. A change of weather, an error of navigation or an attack of sickness and the expedition would now have been doomed. But their luck held. Forsyth reached the Karakash river on the 29th and three days later met a party of Kirghiz with laden yaks sent upstream by the Yarkandis who had safely reached Shahidulla. A few days later the abandoned loads were recovered and the mission proceeded over the Kun Lun without further mishap.

There were, however, other problems. When Forsyth eventually returned to Leh and heard the news of Hayward's death it was not just Kashmiri duplicity that he attacked. He told his brother that he also wished that he had never trusted the word of a Yarkandi. The Kashmiris were out to scare him off. They intended to give him such a horror of the Changchenmo that he would never again recommend it as a trade route. And they succeeded. The Yarkandis had a different game. They wanted the Forsyth mission to reach Eastern Turkestan at all costs, but they also knew that Forsyth's instructions were to proceed only if the country were at peace and if he could expect to conclude his business with Yakub Beg and return before winter. In fact the country was not at peace. Yakub Beg was fighting with malcontents in the furthest corner of his kingdom and there was no hope of Forsyth seeing him before the Kun Lun passes were closed by the snows of winter. But all of this the Yarkand envoy and his men kept to themselves.

At Shahidulla Forsyth started to smell a rat. The Yarkandis continued to insist that Yakub Beg was at Kashgar and the country as a whole at peace. But rumours to the contrary were rife and, once across the Kun Lun, these were amply confirmed. Forsyth determined to return immediately. He didn't like being duped. He shared something of Shaw's dedication, but unlike the diligent planter he was never accommodating. Raised in the Punjab Civil Service under men like the Lawrences, he was used to instant obedience. In matters of state he could be diplomatic enough but it was not inherent in his character. The Yarkandis protested that he couldn't return. He would need more supplies and more ponies and these could only be obtained at Yarkand. Forsyth refused to budge. Only a letter from the governor of the city assuring him that all was well, plus the fact that he was being starved into submission, persuaded him to carry on.

In Yarkand the mission paid for what was now coming to be regarded as Shaw's complacent endurance of the close imprisonment imposed during his previous visit. It was assumed that Forsyth would accept the same treatment. His first attempt to ride round the city caused as much of a furore as had Hayward's. Requests for ponies and provisions for the return journey were met with bland assurances that the king was on his way to Yarkand. It was only by a show of the utmost firmness and a refusal to pay any attention to his hosts, whatever the personal risk or the political consequences, that Forsyth eventually got away. 'I have reason to wish that I had never heard of Yarkand,' he told his brother. 'All hopes of opening out free relations with this country are at an end.' He expected to reach the Punjab by November and 'then farewell to Central Asian affairs for me'.

It was not a promising beginning for Mayo's new policy towards Eastern Turkestan. Forsyth had achieved nothing. In fact his experience of the Changchenmo route had made him doubtful of the feasibility of the whole subject whilst his behaviour in Yarkand had, if anything, made Yakub Beg more suspicious than ever. As for the Russians, they viewed the exercise with intense disapproval. The mission had done as much to precipitate their intervention as to prevent it. They moved more troops to the Kashgar frontier and invasion was only staved off by Yakub Beg agreeing to receive an impressive Russian mission in 1872 which won important commercial privileges.

Only Shaw and Henderson derived any satisfaction from the visit. Henderson secured a specimen of a magnificent falcon, bigger by far than the peregrine, which became known as *Falco Hendersonii* and Shaw, tearing a leaf from Hayward's book, at last won recognition as a genuine explorer. Geographers had been disparaging about his first journey across the mountains. He might claim to be the first Englishman to have reached Yarkand but his account of the journey was deemed far too unscientific and colloquial. Hayward was the explorer; Shaw just a popular propagandist. He encountered something of the distaste that would greet H. M. Stanley three years later. Now, however, he made amends. He had mastered the use of survey instruments and, leaving the Forsyth mission in the Karakorams, set out to make his mark on the map. His objective was the headwaters of the Shyok river. The excursion proved every bit as horrifying as Hayward's to the source of the Yarkand and he duly won his Gold

Medal from the Royal Geographical Society. But the prize cost him dear. Fording one too many ice-swollen rivers his rheumatic fever returned, and by the time he reached Leh he was a very sick man. He never fully recovered. The incredible resilience that saw him doggedly continuing to promote trade with Yarkand when the mission had so obviously failed, and that would even see him crossing the mountains yet again though his constitution was perilously frail, was purely psychological. Though neither colourful nor particularly endearing he was, unquestionably, brave.

14. Back to the Mountains

At two o'clock on a December afternoon in 1873 the citizens of Kashgar turned out in force. It was bitterly cold. They stamped their booted feet on the hard frozen ground and their breath hung in the still air. The pale sun neither coaxed the temperature above freezing nor enlivened the drabness of this brown desert city. Lining the low mud-built walls and huddling deeper into their quilted coats, the down-trodden Kashgaris waited, orderly and silent. It could as well have been another execution that they had come to see. No ripple of excitement had greeted the news that at last a grand mission had arrived from beyond the southern mountains. And there would be no appreciation of the carefully rehearsed protocol of the approaching procession. They were simply curious as, wide-eyed and stony-faced, they watched the cavalcade approach.

In front came two mounted guards, members of the Guides, a crack Indian regiment which specialised in intelligence work. They were part of a detachment seconded to the mission on escort duty. Next, in the place of honour and flanked by two orderlies in scarlet livery with silver sticks of office poised before them, came the Queen's letter. In Calcutta someone had remembered the price paid by Stoddart and Connolly in Bukhara for failing to get the Queen to address the Amir, and someone else had recalled how in the recent campaign in Abyssinia King Theodore had set great store by personal communications from Her Majesty. One could not be too careful with princes so far beyond the reach of civilisation.

Moreover the letter made a worthy centrepiece to the whole presentation. It nestled in a magnificent casket of pale yellow quartz banded with gold and bossed with onyx stones. The handles too were of gold. This rode upon a cushion of rich blue velvet borne at arm's length by the swaggering figure of one Corporal Rhind, 92nd Highlanders. In full dress uniform, silver fittings shining against a dark tartan, his kilt alone was enough to cause a sensation in Eastern Turkestan. The Governor of Yarkand, in fact, had been so taken aback that he had asked whether a mission that hustled out one of its members before he had had time to pull on his trousers was not too obsessed with punctuality.

Behind the Queen's letter came the Viceroy's. It also was in a richly worked casket all red and blue and gold and carried by a scarlet-coated Havildar, the Indian equivalent of a Corporal. Then followed His Excellency the Envoy to the Court of Kashgar and with him Yakub Beg's master of ceremonies. Behind them the members of the mission. All were mounted and in full dress uniform. Swords clanked at their sides, gold lace swayed over the blue and scarlet tunics and cocked hats rode jauntily above the peeled and weather-beaten faces.

There followed the presents, a hundred men bearing silks and cottons from India, guns and gimmicry from England. There were sewing machines, musical boxes (one was actually playing 'Come where my love lies dreaming'), working models of steam engines and steamships, lantern slides and projectors and even a small telegraph outfit; it was hoped that Yakub Beg would allow the mission to transmit messages from their quarters to the palace. Needless to say he didn't. Only the sewing machines caused a sensation; thanks to the Russians their fame had already reached Kashgar. 'The galvanic battery and the wheel of life proved as usual most popular' but the steamships, in a country two thousand miles from the nearest sea and without a navigable river, must have puzzled a few.

By the time the final cavalry escort had made its appearance, the head of the column had reached the palace. The riders dismounted and entered a gate in the unimposing flat-roofed building. They crossed two courtyards, each occupied by four hundred soldiers of the king's bodyguard seated motionless and silent along the walls. At the third gate the master of ceremonies motioned to the party to wait, peeped through and then beckoned them on. They entered another courtyard. This time there was a frozen pool and a few leafless poplars but not a soul was to be seen. The MC with bowed head, folded arms and noiseless step crept towards a raised pavilion and disappeared inside. On tiptoe the British party followed, then waited. 'We took the opportunity to whisper to each other as the tone most suited to the dreadful silence of the spot and the occasion.' It must have roused painful memories of visits to the headmaster.

The Envoy was summoned first. Forsyth—for it was he, in spite of that farewell to Central Asia—entered briskly and was half way to a blank wall on the other side of the room before his host appeared from a side door. Yakub Beg at first sight was unremarkable, middle aged, medium sized, stoutly built, plainly dressed. But once seated on the floor, the civilities over, his face commanded attention. It was full and fleshy, thick-lipped and without a wrinkle—pleasant enough except for the eyes which were cold and hard, neither trusting nor welcoming. It was not a nice expression.

One by one Forsyth introduced his companions. Lieutenant-Colonel T. E. Gordon was second-in-command. The identical twin brother of another Lieutenant-Colonel—they both eventually became generals and were knighted—Gordon was the master of discretion, a quiet, utterly reliable man, well suited to the delicate work of gathering strategic intelligence. Next came Surgeon-Major Bellew, doctor and linguist and a close personal friend of Forsyth. Then Captain Biddulph, an ADC to the Viceroy and a 'political' in the tradition of Burnes. He had already travelled extensively in the Western Himalayas and was soon to be the first British agent in Gilgit. Captain Trotter of the Royal Engineers was the party's surveyor, entrusted among other things with making a dispassionate appraisal of the Changchenmo and Karakoram routes; he had come by the former and was to return by the latter. Captain Chapman of the Quartermaster General's Department was in charge of camp and carriage arrangements. It had been up to him to guard against the sort of disasters that had struck Forsyth's first mission when crossing the Aksai Chin. He had succeeded but thanks largely to a more co-operative attitude from the Kashmiris.

Finally there was the young and gentle man of science, Dr. Ferdinand Stolicska. He was the odd man out, a mid-European civilian in a party of English officers. Forsyth had been inundated with applications to join the mission but few of them can have been from naturalists or geologists. For though Stolicska's qualifications were unimpeachable, he had an unfortunate reputation for guileless naivety. 'I am awfully glad that I have been allowed to go to Yarkand,' was his disarming but somehow improper comment on being appointed. It was ironical too, since he alone never returned to India. He died from the effects of altitude on the Depsang Plateau in 1874.*

* Not far from the bungalow built by W. H. Johnson in Leh stands an obelisk erected by the Government of India to commemorate poor Stolicska. The inscription is still clearly legible.

Under this marble lie the mortal remains of
Ferdinand Stolicska, Ph.D.
born in Moravia, 7th June 1838
died at Malgo, 19th July 1874
while returning from Yarkand with the British mission
etc.

Each man, in a curious mixture of greetings, bowed, shook hands and bid the king 'salaam', then sat on the floor alongside Forsyth. When Rhind's full figure filled the door, all except the king rose to attention. Gordon strode to the door to take the precious burden. Forsyth went half way to receive it from him, turned, advanced to the king and, with one knee bent, placed it before him. The officers closed up in two ranks behind him.

After depositing the casket Mr. Forsyth took out the Queen's letter and, presenting it to the Amir [Yakub Beg's latest title], rose and in a clear sonorous voice, addressed him as follows in Persian.

'I have the honour to present to your highness this letter from Her Most Gracious Majesty the Queen of Great Britain and Ireland and Empress of Hindustan. Since the government of Her Majesty is on terms of amity and friendship with all governments of the world it is hoped that the same relations may be established between the British government and that of Your Majesty'.

The Amir looked very pleased and brightening up said, 'God be praised, you have conferred an honour on me. I am honoured in the receipt of this letter from the Queen. I am highly gratified. God be praised.' And then bowing with the letter in his hands, unrestrainedly enquired, 'Is this box too for me?' And an affirmative reply being given, replaced the letter within it.

The Queen's presents followed and then, with the same formality, the Viceroy's letter and the Viceroy's presents. This time Forsyth's speech was a little more friendly and extravagant. Yakub Beg's countenance betrayed what almost looked like a grin of pleasure. Not, however, his eyes. They remained grim, almost melancholic. Between long silences more platitudes were exchanged till, with a great sense of relief, the procession reformed and made its way back to quarters.

Kashgar was quietly impressed. Forsyth's second mission was not only the grandest ever seen in Eastern Turkestan but the biggest of its kind sent out from India for sixty-five years—since Elphinstone's mission to Kabul in fact. Lord Northbrook, who had succeeded Mayo as Viceroy when the latter was assassinated in 1872, was taking Kashgar seriously. Pressurised by commercial interests at home where the recuperating Shaw had again been active, spurred on by news of more Russian activity in the Tian Shan and by the successful Russian mission to Kashgar in 1872, and approached once again by an envoy from Yakub Beg, he decided to take the plunge. In June he declared firmly that Kashgar was not regarded as within Russia's sphere of influence. And to show that these were not just empty words he inaugurated the sort of mission that would prove that the British considered Eastern Turkestan to be both a vital trading partner and a friendly sovereign neighbour.

To lead the mission, Forsyth was again summoned to Simla for a briefing. Sir Henry Durand once complained that no one in Asia was likely to be fooled by trade missions headed by a man like Forsyth. He was too blatantly a political firebrand. But this was now irrelevant. Northbrook continued to maintain the fiction of seeking only commercial ties beyond the mountains, but he believed in them even less than Mayo. A mission on this scale, with its soldiers and surveyors, was no more dedicated to selling tea than was Forsyth.

The debacle of 1870 had not upset Forsyth's career. He had followed instructions and in the end had managed to keep his disgust out of his official report. He was, however, in trouble. In 1872 he had been responsible for the heavy-handed suppression of a Sikh sect called the Kookas. It was one of those unpleasant incidents which seem to punctuate the normally relaxed flow of Anglo-Indian relations. Forsyth suddenly behaved as if it was the Mutiny all over again. He authorised a number of summary hangings, supported to the hilt his even more blood-thirsty subordinate, and was duly dismissed from his appointment. The Kashgar assignment came as a heaven-sent chance to redeem himself. His manner, according to a contemporary, had become 'lofty and protesting'; others saw his appointment to the mission as a way of doing penance in the wilderness. Forsyth, now in his late forties, saw it for what it was, his last chance.

The mission had formed up under the chenar trees on the edge of Dal lake outside Srinagar. Where Jacquemont had once whiled away the hours with lovelorn dreams of Kashmiri maidens, they rose to the sound of bugles, messed off trestle tables and had Rhind play the pipes after dinner. On September 3rd, they at last got under way. Besides Forsyth, Stolicska and the various army officers the mission consisted of a vast concourse of Indian subordinates. There was the Guides escort, cavalry and infantry, there were secretaries, interpreters, treasurers, medical dispensers, *shikaris* (huntsmen), taxidermists and surveyors and there were all the infinite varieties of camp servants from the cooks to the men who emptied the commodes. The envoy from Kashgar, who overtook them at Shahidulla, had a following of almost equal size including a platoon of Turks recruited in Istanbul.

In retrospect, perhaps one of the greatest achievements of the whole mission was the ease with which it crossed and recrossed the Himalayas. The man who made it possible was none other than William Henry Johnson. The old Kotan pioneer was now governor of Ladakh in place of the Kashmiri who had virtually wrecked the earlier mission. Johnson still loved a *tamasha*, and nothing quite like Forsyth's jamboree had ever hit Ladakh before. The permanent staff

On the Dal Lake

of the mission totalled about three hundred and fifty. Their baggage animals, which were supposed to make them independent of Kashmir help if necessary, ran to some five hundred. As it turned out, they never had to carry a load the whole way to the Kashmir frontier. Johnson not only laid out supplies and fuel at every possible camp-site en route but organised carriage for the whole mission as well as for the additional food needed to feed the mission's own animals.

The economy of Ladakh is said to have taken four years to recover from the unwonted strain. During a period of two months, a grand total of 6,476 porters and 1,621 ponies and yaks plied back and forth between the different stages of the Karakoram route. The same per-formance was repeated for the return journey. Johnson himself went as far as Shahidulla. He felt out the ground ahead, coaxing the mission along and in places taking charge of the whole baggage train so that the envoy, unencumbered, could take a shorter and easier route. It was a magnificent effort and one that, for once in Johnson's unhappy career, did not go unacknowledged. He was congratulated by the government and, when for the first time in his life he visited England, he was awarded a gold watch by the Royal Geographical Society.

When he left the mission in Shahidulla their beasts were as fresh as they had been in Srinagar. Had they not been, Forsyth would have been pushed to cross the Kun Lun in safety. As it was there was a serious scare on the Sanju where ice and avalanches threw the whole column into confusion. Twenty ponies were killed and a number of benighted followers lost toes and fingers through frostbite. It all went to show with what skill and care they had been piloted across the far worse Sasser and Karakoram. For once no one suggested that the Leh–Yarkand road was a feasible invasion route. For trade purposes Forsyth and Trotter agreed that the Karakoram was preferable to the Changchenmo but, without the co-operation of a man like Johnson, neither was capable of supporting a modest force of infantry let alone artillery.

The size of the mission was dictated by its almost unlimited field of enquiry. Minute examination was to be given to every aspect of life in Turkestan and the final report, though considerably less readable, puts Alexander Cunningham's work on Ladakh firmly in the shade. The history, literature, antiquities, languages and ethnology of the country were explored by the two scholars, Bellew and Biddulph. To them we owe such interesting details as the easiest way to get a divorce in Eastern Turkestan—for the woman, evidence of unnatural behaviour by her husband could be delicately indicated by her placing her shoes upside down in front of the judge. Exhaustive details of the retail and wholesale trade were collected by Chapman who plunged into the bazaars and emerged with some novel descriptions of how the Kashgaris dealt with the natural incontinence of their babies. From every swaddled infant projected a skilfully designed tube procurable from shops devoted solely to the sale of this item. Gordon's job was to cast a critical eye over the rabble of Khokandi, Kirghiz and Chinese troops which composed Yakub Beg's army. He was not much impressed. Bows and arrows were still used by some contingents, and the formations of converted Chinese relied on a musket some nine feet long which it took two men to fire and another two to hold. Busiest of all was Stolicska with his native plant gatherers and bird stuffers. He witnessed the flying of eagles against foxes and wolves, which was very much a speciality of the country, and he amassed an enormous collection of rocks, plants and animals, including the wild sheep of the Tian Shan, a close relative of the Marco Polo sheep of the Pamirs.

All of which was in addition to the main task of the mission. This inevitably was commercial and political, geographic and strategic. On paper, that is in the readily understandable and comparatively uncontroversial terms of trade, the exercise was a dazzling success. A treaty was concluded with Yakub Beg which gave goods from India unrestricted entry at a low duty of two and a half per cent. Commercial agents could be appointed to places like Kotan and Yarkand, and a British representative with the status of ambassador could be installed in Kashgar. This was a lot more than the Russians had obtained in 1872. Provided he had a passport, any British subject could now enter Eastern Turkestan. Any amount of tea and cutlery and fabrics could be sent across the Himalayas and, through the ambassador in Kashgar, there was an easy avenue for developing closer political ties as the situation dictated. It looked as if British influence in Kashgar was to be even stronger than in Kabul.

When Forsyth returned to India the following year his fame was assured. He was greeted as a hero. The mission was hailed as the greatest diplomatic success for forty years. Forsyth, rebuffed for so long by Lawrence, castigated by Mayo over the Kooka affair and ridiculed by the press as late as 1873, was vindicated. He was instantly knighted and most of his companions took a step up in rank.

But in this, their hour of glory, Forsyth, Bellew and the rest had their reservations. There was room for expansion of the Yarkand trade but they could not see that it was ever likely to amount to much. Shaw, from a mixture of ignorance and over-enthusiasm, had got it all wrong. The place was largely desert, the people didn't like Indian tea, Chinese goods including the traditional tea were already finding their way into the country and, anyway, the mountains would always have the last word as a limiting factor. They had seen far more of the country than any previous visitor and it was no Eldorado.

They had also gained a far better insight into the character of Yakub Beg and his rule. This again gave cause for concern. Crime was almost unheard of, the people were orderly, the roads safe and the government secure. But by no stretch of the imagination could Yakub Beg be regarded as an enlightened ruler. He was a despot, as cruel and unprincipled as the worst of his equivalents in Western Turkestan. The country was ruled by fear and oppression. Those, including Queen Victoria, who had a sneaking admiration for this new bastion of resurgent Mohammedanism in Central Asia, would do well to revise their ideas. Here was no grafting Tudor moulding a new identity and unity for his people but a grim, unloved Cromwell. 'One of the most notable features of peasant life', wrote Bellew, 'was its eerie silence . . . we never heard the sound of music, the voice of song or the laugh of joy', and the reason was the repressive character of the government. For all its orthodoxy Yakub Beg's rule was detested by most of the people. They looked forward to the day when the Russians would intervene or the Chinese be restored. When Forsyth's mission withdrew, long faces and whispered exchanges hinted at the imminent revival of the reign of terror which had been temporarily suspended. The gigantic gallows, dismantled for the period of their stay, were about to be re-erected.

As soon as the mission returned to Kashmir, Robert Shaw, apparently recovered and now Joint Commissioner in Leh (he had given up his tea estates in favour of government employ), was packed off to Kashgar to get the treaty ratified and to establish himself as the new ambassador. He failed on both counts. Yakub Beg refused to allow him the freedom of movement enjoyed by the mission, and he found himself in much the same situation as in 1868. During a stay of six months he was never even permitted to enter the main city of Kashgar. Perhaps he had been the wrong man to send; he had always been too amenable. But Northbrook also was having second thoughts. Yakub Beg obviously did not want an ambassador, and in London there were fears that Shaw might be creating more ill-will than good. They also feared for his safety and felt that, if the Russians did eventually move in, he could only prove an embarrassment. He was duly recalled in 1875. He brought with him what he imagined to be a fully ratified copy of Forsyth's treaty. But, on examination, Yakub Beg's seal turned out to be just a complimentary note to the Viceroy. Unratified, the whole thing was worthless.

Vastly more significant was the geographical work of the mission. Forsyth planned to visit all the major cities of Eastern Turkestan, to explore the Tian Shan frontier with Russia and to return to India via the Pamirs and Afghanistan. Yakub Beg, of course, resisted. The British might, with a guide, go wherever they wanted within a day's march of their headquarters but expeditions beyond this required his specific approval. In practice, Kotan was out. So too was Aksu in the north-east, though Biddulph was allowed to visit Maralbashi, a city half way towards Aksu.

Rather surprisingly Gordon, Trotter and Stolicska were allowed into the Tian Shan. They reached one of the main passes just north of Kashgar and stood within thirty miles of the Russian fort on the Naryn river. It was as near as a British officer on duty had been to the Russian frontier for thirty years, by the rules of the Great Game a mighty move. An equivalent would have been that of a party of Russian officers surveying their way across Afghanistan and peering inquisitively down the Khyber pass. Imagine the pandemonium this would have caused; the outcry in St. Petersberg was remarkably restrained by comparison. In the process they also carried the survey from India across ground already covered by the surveyors from Russia. On the maps at least, the two empires had met.

Gordon surmised that it was his exemplary tact on this foray into the Tian Shan which finally persuaded Yakub Beg to agree to the still more ambitious scheme of crossing the Pamirs. In fact his permission was revoked; but too late. Gordon, with Biddulph, Trotter and Stolicska, had leapt at the chance and, in spite of the unfavourable season, had left on March 21st 1874. Forsyth had been warned that the Afghans would probably object to the whole mission returning via the Pamirs and Kabul, but Gordon was to verify this. He was also to enquire further into the circumstances of Hayward's murder, to clear up the question of the source of the Oxus and to explore as much of the region as possible in view of its growing strategic importance. Since Lieutenant John Wood's lonely odyssey in the winter of 1838, no European had seen the Roof of the World. Adolph Schlagintweit had probably hoped to cross it and he was murdered at Kashgar. Hayward had tried and he was killed in Yasin. A few native agents had been there, the Mirza in 1869 and a man sent by Forsyth in 1870, but the only Europeans, apart from Wood and perhaps Gardiner, were still Marco Polo and Benedict de Goes. To the geographers in Europe a first-hand account of its hydrography and mountain structure was the great 'desideratum'—their favourite word—of Asian geography.

Gordon and his companions went some way towards providing it. The story of their journey belongs to a later phase of Himalayan exploration and one in which the impetus to find a way through the mountains from India to Central Asia plays no significant part. It is sufficient here to notice just the results. They went by way of Tash-kurgan and the Little Pamir as far as Wakhan where they learnt that Afghanistan was indeed still closed to travellers and, in view of the strained state of Anglo-Afghan relations at the time, particularly so to official missions. As for Hayward's murder, the Mir of Wakhan had his own theories which, not surprisingly, implicated his old enemy Aman-ul-Mulk of Chitral. This was not exactly news.

What really mattered was that at last someone had filled in the gap between Wood's exploration up the Oxus and what was now known of Eastern Turkestan. The circuit of the northern frontier had been completed, and the expedition actually returned from Wakhan to Yarkand by way of Wood's old route up to Lake Sir-i-kol and then on across the Great Pamir. They even met an old man who remembered Wood and was delighted to meet fellow countrymen of the shy young man who had come amongst them back in 1838. Gordon agreed that Sir-i-kol, though it seemed a different shape and was not quite as high as Wood thought, was definitely one of the Oxus sources. He also reckoned it was a lot deeper than Wood's efforts with the plumb-line had indicated and, since Sir-i-kol was not really a proper name, he revived Wood's suggestion of Lake Victoria.

He left alone the question of whether it or the Sarhad was the main feeder since he had found, flowing from a lake on the Little Pamir, another entirely different Oxus tributary (the Aksu, Bartang or Murghab) which he believed to be the true parent river. If it was, then there was good cause for concern. London and St. Petersberg had just agreed on a definition of Afghan territory in the region by which the Oxus, from Wood's lake west, was to be the northern frontier. Rivers seldom make good boundaries and, as Gordon observed, most of the Oxus states, like Wakhan, stretched along both sides of the river. But, apart from this, if the Oxus did not begin anywhere near Wood's lake, then the whole agreement was meaningless.

Nor was this the worst of the expedition's discoveries. Strategi-cally the Pamirs had always been written off as too bleak and barren to appeal to the Russians and too formidable for them to cross. Wood's story suggested nightmarish conditions, which the Mirza's travels fully supported. And, in April, Gordon found the going quite as bad, the wind unbearable, the snow freezing to their faces as it fell, and fuel and provisions desperately short. But from the Wakhi people he heard a different story. In summer the grazing was, as Marco Polo had recorded, some of the best in the world. Moreover, though mountainous, it was nothing compared to the Karakorams or the Kun Lun. The Pamirs, he was told, 'have a thousand roads'. With a guide you could go anywhere and, in summer, considerable forces might cross without difficulty. The Chinese had done it in the past, the Wakhis had recently sent a contingent across to Kashgar, and the Russians might do it in the future.

Finally, and most important of all, it was discovered that the passes leading south from the Pamirs over the Hindu Kush to Chitral, Gilgit and Kashmir were insignificant. This was so disconcerting that Gordon, ever discreet, omitted all mention of it in his published account. The discovery was made by Biddulph who, while the others explored the Great Pamir, made a 'lonely journey by the Little Pamir' (a misprint in Gordon's book actually has it as a 'lovely journey'). In the process he climbed the northern slopes of the Hindu Kush and ascertained that at least two passes constituted veritable breaks in the mountain chain. One you could ride over without ever slowing from a gallop and both had had artillery transported across them. To these Gordon added yet another 'easy pass' conducting from Tashkurgan to Hunza.

In effect, Forsyth's mission had not so much thrown open Eastern Turkestan as refocused attention on the Western Himalayas. The possibility of a Russian invasion via Kashgar into Ladakh could be discounted. If the Russians did eventually take Eastern Turkestan, that country would serve not as a springboard but as a secure flank from which to support a direct thrust across the Pamirs and the Hindu Kush. In that direction there was no intelligible Afghan frontier for them to worry about, the Roof of the World was no impassable wilderness, and India's mountain wall seemed to have a gaping hole.

It was some paradox that for fifty years all eyes had been on those spine-chilling efforts to scale the Karakorams and the Kun Lun in order to reach Central Asia, when a few hundred miles to the west the traveller could, in Biddulph's words, 'go through a gate by which . . . he is practically landed in Central Asia in a single march'.

Biddulph later had cause to moderate his language. The new 'key to India' did not turn quite so easily. In the Gilgit and Chitral regions factors other than the height of the passes had slowed penetration and would continue to do so. The odd thing was that it had taken years of conquering the Western Himalayas to establish that they were not a simple barrier. The explorers themselves knew this. 'Range upon range of mountains meets the eye', 'a chaos of mountains unending', 'a sea of peaks without system or limit'. In Ladakh and Baltistan the surveyors had actually proved it. Yet, for the statesmen and strategists, the truth only now began to sink in. It was not enough to force a path through the mountains at one point and assume that the rest of the region was equally formidable. The Western Himalayas still needed to be systematically explored.

In words curiously reminiscent of those used by H. H. Wilson in his introduction to Moorcroft's *Travels* nearly forty years before, Biddulph summed up the situation.

Anyone who considers the question cannot fail to be struck by the ignorance which prevails concerning the countries immediately beyond our border. It is hardly an exaggeration to say that we know more of the geography of Central Africa than of the countries lying at our own door.

The frontier had advanced a good bit since Moorcroft's day, but Biddulph makes it plain that he is writing not of Central Asia but of Kashmir. Her boundaries were still undefined because the government only now began to realise that they were dealing not with a straightforward range but a vast, unbelievably complex mountain knot.

*   *   *

William Johnson died, probably of poison, in 1878. Shaw, whose delicate constitution had never fully recovered from the strain of his return journey in 1870, was soon moved to Burma and died there in 1879. Forsyth also had a brief spell in Burma and retired to England in 1876. None was therefore on the spot when the policy on which they had staked so much was suddenly and unexpectedly undermined. For in 1877 Yakub Beg was murdered, and within a year the Chinese were back in Eastern Turkestan.

This was the one eventuality that no one in India had foreseen. The country's independence had been thought vulnerable from the north, not the east; the Chinese had long ago been written off as quite incapable of mounting an offensive so far from home. Yet, against a disarrayed enemy and a far from hostile civilian population, they deployed an impressive army which reoccupied the Land of the Six Cities with scarcely a shot being fired. To emphasise the once-

and-for-all character of their reconquest they renamed the country Sinkiang, 'The New Dominion'. The Russians had always managed to hedge their bets on Yakub Beg. They had withheld recognition of him for as long as possible and they had ensured that, if the Chinese should return, they held bargaining counters with which to trade for continued links with Kashgar. Not so the British. They had backed a loser—in Chinese eyes a traitor—and the trade between Leh and Yarkand now plunged into a decline. The Russians installed a fully fledged and immensely influential consul in Kashgar, but no British official was allowed to reside there until the 1890s. It was indicative of the waning significance of the Leh–Yarkand route that, when a British consulate was finally established, its incumbents, rather than chance their luck on the Karakorams, invariably got there via Gilgit.

But a more fitting and significant postscript to a story so enriched by intrepid individuals and so full of misplaced endeavours and tragic consequences was yet to come. It happened on the 8th April 1888. That night on the very crest of the Karakoram pass, a lone Scotsman was hacked to death. His name was Andrew Dalgleish and at the time he was heading for Yarkand with a caravan of piece goods. April was no month to be crossing the mountains but he knew the track and conditions as well as anyone. In fact, for the last fourteen years his livelihood had depended on them.

The snow lay deep on the pass and, at an altitude of close on 19,000 feet, the last steep pull proved punishing. Dalgleish and his men—two servants and seven pony men—decided to halt. They were joined by a Yarkandi trader with a small caravan who was travelling by the same stages. There were also two Afghans and an itinerant fakir. All were known to Dalgleish and, in the course of the evening, he wandered over to their tent. He was offered tea but declined. However he accepted a piece of bread, sat down and they talked for a while. Then one of the Afghans slipped out. Dalgleish asked where he was going. The man said it was a call of nature and disappeared into the night.

Moments later a shot was fired. At point blank range, it passed through the tent and into Dalgleish's shoulder. He stumbled out into the dark heading for his own tent and his gun, but the Afghan was waiting for him. They were ill-matched. Daud Mohammed was a giant of a man and he wielded a massive sword. Dalgleish was short and thin, badly wounded and without so much as a knife. He tried to close on his enemy and actually grabbed the blade of his sword.

But eventually he fell. The blade at last cut through his thick sheep-skins and Daud Mohammed hacked away till he moved no more. No one came to his rescue. Only his terrier joined in the fray and soon suffered the same fate. Dog and master were left, a crumpled mess in the blood-spattered snow.*

Daud Mohammed had known Dalgleish for many years. Though an Afghan, he had been born in what was now British India and in Yarkand had been part of the expatriot Indian community to which Dalgleish also belonged. He was a braggart and a bankrupt but just why he had suddenly taken it into his head to kill Dalgleish was not clear. It could have been just a case of robbery with violence, or it could have been an outburst of religious fanaticism. More probably he was acting on instructions; but who had hired him? A commercial rival perhaps, the Chinese or even the Russians? All these were possible, but unless the murderer could be arrested the truth must remain a mystery.

In spite of protestations from the British, Daud Mohammed passed scot-free through Yarkand and Kashgar and then disappeared. In a final bid to bring him to book Hamilton Bower, a British officer who happened to be in Sinkiang combining a hunting trip with a bit of clandestine surveying, was invited to take up the trail. It was two years since the murder. Daud Mohammed was variously rumoured to have left for Mongolia, Mecca, Kabul and Bukhara. Bower under-standably thought the whole thing hopeless but, in between further hunting forays, he despatched men in all four directions. He was as amazed as anyone when, by an extraordinary coincidence, one of them came face to face with the murderer in a bazaar in Samarkand. The Russians duly arrested him and it looked as if the mystery would at last be cleared up. But Daud Mohammed thought other-wise. According to the Russians he took his own life.

Dalgleish's murder thus remains as unexplained as the deaths of Moorcroft and Trebeck or the slaughter of Hayward. But what makes it particularly relevant in the present instance is the point that Dalgleish was the last and, after Shaw, the only Briton ever to be actively engaged in the Leh–Yarkand trade. Hard on the heels of Forsyth's second mission, he had reached Yarkand in 1874 with the first consignment of goods from the Central Asian Trading Company,

---

* Dalgleish's body was recovered before it became unrecognisable from the other skeletons that strewed the Karakoram route. His grave is beside that of Stolicska in Leh.

a venture promoted by Shaw. He had stayed on in Yarkand as agent of the company then, when it failed, as liquidator, and had finally set up in business on his own. For years he was the only British resident in Sinkiang. He came to know, more intimately than any of his countrymen before or since, the horrors of what was now recognised as 'the worst trade route in the whole wide world' and the problems of what was surely the most fraught commercial venture of the century. It was with his death that the sanguine expectations of William Moorcroft and the exaggerated claims of Robert Shaw were finally laid to rest.

A small white marble pillar was erected in Dalgleish's memory. The inscription was very simple; 'Here fell A. Dalgleish, murdered by an Afghan.' It was placed on a cairn beside the crest of the Karakoram pass and there, on the grim windswept watershed between the rivers of India and Central Asia, at the gateway to which were directed the aspirations of all the early explorers of the Western Himalayas, it may still stand. So long as the present-day ceasefire lines between China, India and Pakistan all meet on that very spot, it would be hard to prove otherwise.

'The Road to Moscow'

Bibliography

I CHAPTER NOTES

Chapters 1 and 2

It is a brave man who sits down to reconstruct Moorcroft's travels from the India Office Library's collection of Moorcroft Manuscripts. I have referred to them only where H. H. Wilson's edition of *The Travels of William Moorcroft and George Trebeck* (London 1841) seems to raise doubts, or in order to pursue specific lines of enquiry. Other unpublished sources to which reference has been made include Trebeck's letters to George Leeson and Stirling's account of the fate of the expedition as given by Askar Ali Khan ('Luskeree Khan') both in the Royal Geographical Society's archives and relevant volumes of Bengal Political Consultations in the India Office Records. John Murray has also kindly supplied copies of correspondence with H. H. Wilson.

Moorcroft's 1812 journey is described in Asiatic Researches vol XXI (Calcutta 1816). Francis Watson's article in the Geographical Magazine (vol XXXIII 1959) also deals with this journey.

The Asiatic Journal (vols XIX old series to XXI new series, 1825–36) has frequent references to Moorcroft's last journey, including a letter from Bukhara and Ghulam Hyder Khan's account of the journey as far as Peshawar.

The Journal of the Royal Geographical Society (vols I and II 1831 and 1832) carries Moorcroft's notes on Kotan and Kashmir.

Of more recent works the following have been particularly useful:

H. W. C. Davis' *The Great Game in Asia 1800–44*, London 1926
G. J. Alder's *British India's Northern Frontier 1865–95*, London 1963
A. Lamb's *British and Chinese Central Asia 1767–1903*, London 1960
K. Mason's *Abode of Snow*, London 1955
S. Hedin's *Southern Tibet* vol VII, Stockholm 1922.
and the introduction to the Moorcroft MSS in the *Catalogue of the India Office Library* (Manuscripts in European Languages vol II).

Chapter 3

The standard biography of Jacquemont is by Pierre Maes, *Victor Jacquemont*,

Paris 1934. David Stacton's *A Ride on a Tiger*, London 1954, gives an entertaining account of his period in India.

His journal was published in six volumes entitled *Voyage dans l'Inde pendant les années 1828–32*, Paris 1835–44, but more entertaining are the letters. These, as Prosper Mérimée says, tell the tale better than anyone could. There are several collections. I have used:

1. *Correspondance de Victor Jacquemont avec sa famille et ses amis pendant son voyage dans l'Inde 1828–32*, New Edition, Paris 1869.
2. *Letters from India*, translated from the French, London 1834.
3. *Letters from India 1829–32*, translated and with an introduction by C. A. Phillips, London 1936.

In this and subsequent chapters, descriptions of Ranjit Singh and his court have been taken at random from the accounts of travellers like Burnes, Vigne and von Hugel and from those of a motley collection of other visitors and residents. These include:

Up the Country, The Hon. Emily Eden, London 1866.

The Court and Camp of Runjeet Singh, The Hon. W. G. Osborne, London 1840.

Events at the Court of Ranjit Singh, Punjab Govt. Record Office, Lahore 1935.

Travels in India, L. von Orlich, London 1845.

Five Years in India, H. E. Fane, London 1842.

The Punjaub, Lt. Col. Steinbach, London 1845.

Journal of a March from Delhi to Peshawar and Kabul, W. Barr, London 1844.

Thirty-Five Years in the East, J. M. Honigberger, London 1852.

Chapter 4

Wolff's autobiography, *Travels and Adventures*, London 1860, sometimes contradicts and invariably elaborates on his journals. The relevant journal is *Researches and Missionary Labours among the Jews, Mohammedans and other sects*, London 1835.

He apparently dictated the autobiography to relays of secretaries with little or no reference to the journals. This suggests a prodigious memory and it is remarkable that there are not more inconsistencies.

There is a biography of Wolff—*Joseph Wolff, His Romantic Life and Travels*, by H. Palmer, London 1935. It is compiled almost exclusively from the *Travels and Adventures* and, considering the subject, is surprisingly dull.

His 1844 journey to Bukhara to discover the fate of two British officers,

Stoddart and Connolly, has been well told by Fitzroy MacLean in *A Person from England and other Travellers*, London 1958. The journal for this expedition has recently been reprinted as *A Mission to Bokhara*, edited and with introduction by Guy Wint, London 1969.

In the Royal Geographical Society Archives there are a few letters from Wolff. They are chiefly concerned with his being unable to afford the subscription; he needed the money to buy coal for his poor parishioners.

Chapter 5

Professor H. W. C. Davies in his Raleigh Lecture to the British Academy in 1926, *The Great Game in Asia, 1800–1844*, also implies that Vigne's travels had reference to Anglo-Russian relations. He claims, however, that Vigne met Henderson in Ladakh and that part of his mission was to convince Ahmed Shah that 'not Russia but Great Britain . . . would be the truer friend'. Vigne had met Henderson before but in Ludhiana, not Ladakh, and I have found nothing to suggest that Ahmed Shah needed any convincing about where his best chance of salvation lay. But that Vigne went to Baltistan 'to spy out the land' seems highly probable. Vol IV of R. H. Phillimore's *Historical Records of the Survey of India* (1945–65) has been useful in this connection.

Henderson was again robbed of everything, including his diary, on his last journey and died before he was able to write even a summary of his travels. The relevant works of G. T. Vigne are *Travels in Kashmir, Ladakh, Iskardo etc.*, 1842, and *A Personal Narrative of a Visit to Ghazni, Kabul and Afghanistan*, 1840.

Von Hugel's work was translated into English as *Travels in Kashmir and the Punjab*, 1845. The Journal of the Asiatic Society of Bengal (JASB) vol V 1836 and vol VI, part II, 1837, The Asiatic Journal vol XXI (New Series) Sept–Dec 1836 and the Journal of the Royal Geographical Society (JRGS) vol IX 1839 contain letters and notes by Vigne. The latter (JRGS) vol VI 1836 includes a contribution from von Hugel. Their books are reviewed in vols XI 1841, XII 1842 and XXXI 1861, presentation of von Hugel's medal is in vol XIX 1849 and his obituary in Proceedings of the RGS (PRGS) vol XV 1870–71.

For biographical information on Vigne I am indebted to Mr. H. D'O. Vigne and to the entry in the *Dictionary of National Biography*.

Sven Hedin's *Southern Tibet* vol VII and Kenneth Mason's *Abode of Snow* along with JRGS vol XXXI 1861 have the best assessments of Vigne's geographical work.

Chapters 6 and 7

So many have succumbed while working on Gardiner's story that one moves on with a sigh of relief. Since Major H. Pearse edited his *Memoirs of Alexander Gardiner*, London 1898, the lethal Gardiner papers, including Cooper's draft, have again disappeared. Should they ever come to light, it will be a brave man who undertakes to re-examine them. I doubt too whether they will help to solve the mystery.

For Gardiner's spell in the Sikh service Pearse used them hardly at all, but it is on evidence from this period that C. Grey in *European Adventurers of Northern India 1785–1849*, edited by H. L. O. Garrett, Lahore 1929, bases his demolition of the case for Gardiner.

Sir H. M. Durand's *Life of a Soldier of the Olden Time* will be found in *Life of Sir H. M. Durand*, by H. M. Durand, vol II, London 1883. Edited extracts from Gardiner's journals are in Journal of the Asiatic Society of Bengal (JASB) vol XXII, 1853 and information derived from his employment at Lahore in *A History of the Reigning Family of Lahore*, edited by G. Carmichael Smyth, Calcutta 1849.

The minute dealing with Gardiner's gun-running operation is dated June 30th 1874 and is in Secret Home Correspondence, vol 79 in the India Office Records.

Other works referred to in this chapter are as follows:

C. Masson's *Narrative of Various Journeys in Balochistan, Afghanistan and the Punjab*, London 1842.

T. E. Gordon's *A Varied Life*, London 1906.

G. W. Leitner's *Languages and Races of Dardistan*, Lahore 1877.

A. Wilson's *Abode of Snow*, Edinburgh 1875.

G. Hayward's letters in Proceedings of the RGS (PRGS) vol XV 1870–71.

Edinburgh Review, January 1872.

Journal of the RGS vol XLII 1872.

Chapters 8 and 9

Besides *Travels into Bokhara*, London 1834, Sir Alexander Burnes wrote *Cabool, A Personal Narrative of a Journey to and Residence in that City, 1836, 7 and 8*, which was published posthumously in 1842. The Bukhara journey is also dealt with in *Travels in the Punjab etc.*, London 1846, by Mohan Lal, Burnes's interpreter, and in the letters of Dr. Gerard published in JASB vol II 1833. Charles Masson's *Narrative of Various Journeys in Balochistan, Afghanistan and the Punjab etc.*, London 1842, gives a valuable insight into the events of 1836–38 in Kabul and Vigne's summary in *Travels in Kashmir etc.* also provides a useful, because impartial, assessment. Burnes's letters were published as Parliamentary Papers and are quoted in more digestible form in an article by George Buist of the *Bombay Times* which may be found in a little book entitled *Notes on his Name and Family*, by James Burnes, Edinburgh 1851. James Lunt in his recent biography, *Bokhara Burnes* (edited by George Woodcock), London 1969, has also used the Buist article extensively.

Burnes's role in the Great Game is discussed by H. W. C. Davies in *The Great Game in Asia*, J. L. Morison in *From Alexander Burnes to Frederick Roberts* and T. H. Holdich in *The Gates of India*.

The only unpublished papers to which I have had recourse are the Reports of the Burnes Mission to Kabul, 1837-8, in the India Office Library and some letters of Burnes to John Murray.

Wood is an altogether more elusive subject. His *Journey to the Source of the River Oxus* was published in 1841. A second edition, with a preface by his son and a long introduction by Sir Henry Yule, appeared in 1872. The description of the journey in JRGS vol X 1840 adds nothing to the book. His only other published work seems to have been a pamphlet accusing the New Zealand Company of misrepresenting the delights of their islands. PRGS vol XVI 1871-2 contains Rawlinson's obituary notice and from this, together with his son's preface as above and the entries in *DNB* and in vol IV of R. H. Phillimore's *Historical Records of the Survey of India*, I have ventured to reconstruct something of his character and career. There are a few unrevealing letters from him in the RGS archives but sadly no portrait.

Chapter 10

The quotation from Sir William Lloyd comes from *Narrative of a Journey from Caunpoor to the Boorendo Pass*, Major Sir William Lloyd, London 1840. He also has a magnificent description of sunset over the mountains seen from Kotgarh. For the history of Simla see *Simla, Past and Present*, by E. J. Buck, Bombay 1925.

The Tibetan Campaign of Zorawar Singh is described in Cunningham's *Ladakh* and is discussed in K. M. Pannikar's *The Founding of the Kashmir State; a Biography of Gulab Singh*, London 1953, and in *Britain and Chinese Central Asia, The Road to Lhasa*, by Alistair Lamb, London 1960. I have also used the latter for its lucid presentation of the background to the 1847 Boundary Commission.

The official instructions to the Commission are in vol III no. 48 of Enclosures to Secret Letters in the India Office Records. The reports of the Commission are in Board's Collections vol 2461 no. 136806, 1851.

JRGS vols XVIII and XIX 1848 and 1849, contain the first mentions of the 'expedition' and vols XIX and XXIII contain the abbreviated accounts of Thomson and Strachey. Letters from Cunningham and Thomson were published in JASB vol XVII 1848. The three great productions of the Commission are *Ladak, Physical, Statistical and Historical*, by Alexander Cunningham, London 1854, *The Physical Geography of Western Tibet*, by Henry Strachey, London 1853, and *Western Himalaya and Tibet*, by Thomas Thomson, London 1853.

Sven Hedin's *Southern Tibet* vol VII and Kenneth Mason's *Abode of Snow* contain geographical appraisals of the Commission's work.

Chapter 11

For a recent publication R. H. Phillimore's *Historical Records of the Survey of India*, 1945–1965, Dehra Dun, is hard to come by. It is a mine of information on the Kashmir survey and well worth tracking down. Shorter accounts may be found in *A Memoir of the Indian Surveys*, by Clements R. Markham, London 1871, and in Kenneth Mason's *Abode of Snow*; longer ones in the annual *General Reports of the G.T.S.*, in the *Records of the Survey of India* and in the synoptical volumes of the same.

Godwin-Austen's report *On the Glaciers of the Mustakh Range* was published in JRGS vol XXXIV 1864 and Johnson's *Report of his journey to Ilchi, the capital of Khoten* in JRGS vol XXXVII 1867. Discussion of the first will be found in PRGS vol VII 1863–4 and of the second in PRGS vol XI, 1866–7. Johnson's obituary is in PRGS vol V 1883.

Godwin-Austen's comments on Montgomerie are to be found in a letter to Bates of the RGS in the RGS archives. For the other side of the story see Everest's letter to Norton Shaw of May 11th 1861 also in the RGS archives. The story of K2 comes principally from Phillimore as above.

The repercussions of Johnson's visit to Kotan can be followed at length in vol 91 no. 93 of Collections to Political dispatches to India in the India Office Records. They are discussed by Alistair Lamb in *The China–India Border*, London 1964.

Gluttons for punishment may reconstruct the travels of the Schlagintweit brothers from *Results of a Scientific Mission to India and High Asia*, London 1861–5, *Reisen in Indien und Hoch Asien*, Jena 1880, PRGS vol I 1857 and JRGS vols XXVII 1857 and XXVIII 1858. Sven Hedin in *Southern Tibet* vol VII summarises their work and the British attitude to them may be gathered from the delightful *Travels in Ladakh, Tartary and Kashmir*, by H. D. Torrens, London 1862, and from letters in the RGS archives from George Buist (14 Nov. 1854) and George Everest (Nov. 1861).

Chapters 12, 13 and 14

Accounts of the first British journeys into Eastern Turkestan are by Robert Shaw, *Visits to High Tartary, Yarkand and Kashgar*, London 1871, George Henderson and Allan Octavian Hume (ornithologist and founder of the Indian Congress Party), *Lahore to Yarkand*, London 1873, H. W. Bellew, *Kashmir and Kashgar*, London 1875, T. E. Gordon, *The Roof of the World*, Edinburgh 1876, and *Sir Douglas Forsyth*, an autobiography edited by Lady Forsyth, London 1887.

These together with the official *Report of a Mission to Yarkand and Kashgar in 1873* were raided by several writers who cashed in on the Eastern Turkestan fad during the 1870s. They included D. C. Boulger (*Yakoob Beg*, London 1878) and G. R. Alberigh-Mackay (*Notes on Western Turkestan*, Calcutta 1875).

Hayward's activities are chronicled in the Journal and Proceedings of the Royal Geographical Society 1868–71. His report on the journey to the source of the Yarkand river and to Eastern Turkestan is in JRGS vol XV 1870; vol XVI 1871 contains Montgomerie's report on the Mirza's journey. The Society's publications give much additional information on the Forsyth missions and on Shaw's journey down the Shyok in 1870. Reading these one gets a good idea of the excitement which attached to the whole subject of exploration in the Western Himalayas. Only Dr. Livingstone's disappearance and death are allowed to share the limelight.

There are a few political memoirs which bear on the subject, notably O. T. Burne's *Memories*, London 1907, G. R. Elsmie's *Thirty-five Years in the Punjab*, Edinburgh 1908, the *Life and Letters of H. M. Durand*, by his son Sir H. M. Durand, London 1883, and *A Varied Life*, by T. E. Gordon, London 1906.

An exhaustive study of the political background to the various journeys will be found in G. J. Alder's *British India's Northern Frontier 1865–95*. This makes light work of an extremely complicated subject and I owe Dr. Alder a great debt of thanks. His article on Hayward's murder in Journal of Royal Central Asian Society, 1965, has also been consulted.

Letters from Shaw and Hayward in the RGS archives have proved useful in tracing their changing attitudes towards Yakub Beg and their relationship with one another. Some of Hayward's watercolours are filed with them. Hayward's murder is the subject of endless reports and correspondence in the India Office Records. These shed little light on the man himself but a lot on the attitudes of Forsyth, Cayley and others towards the Maharaja of Kashmir. Reports by Gordon and Biddulph on the Pamirs expedition of 1874 are also in the India Office Library.

The account of Dalgleish's murder has been taken chiefly from H. Bower's *A Trip to Turkestan* in the Geographical Journal vol V 1875 and H. Lansdell's *Chinese Central Asia*, London 1893.

II GENERAL BIBLIOGRAPHY

For the most part this list is supplementary to the works mentioned in the Chapter Sources and consists of recent publications or those relevant to the subject as a whole.

ALDER, G. J. *British India's Northern Frontier 1865-95*, London 1963.

CURZON, G. N. *The Pamirs and The Source of the Oxus*, London 1896.

—— *Russia in Central Asia*, London 1889.

DABBS, J. A. *Discovery of Chinese Turkestan*, The Hague 1963.

DAINELLI, G. *La Esplorazione della Regione fra L'Himalaja Occidentale e il Caracorum*, Bologna 1934.

DAVIS, H. W. C. *The Great Game in Asia 1800-44*, London 1926.

DREW, F. *Jummoo and Kashmir*, London 1875.

EDWARDES, M. *Playing the Great Game*, London 1975.

FAIRLEY, J. *The Lion River; The Indus*, London 1975.

HEDIN, S. *Southern Tibet*, Stockholm 1922.

HOLDICH, T. H. *Indian Borderland*, London 1901.

—— *Gates of India*, London 1910.

LAMB, A. *British and Chinese Central Asia*, London 1960.

MASON, K. *Abode of Snow*, London 1955.

MORRISON, J. L. *From Burnes to Roberts*, London 1936.

PHILLIMORE, R. H. *Historical Records of the Survey of India*, Dehra Dun 1945.

RAWLINSON, H. *England and Russia in the East*, London 1875.

SEVERIN, T. *The Oriental Adventures: Explorers of the East*, London 1976.

WILSON, A. *Abode of Snow*, London 1876.

Index